EAST LANCASHIRE
MINING MEMORIES

EAST LANCASHIRE MINING MEMORIES

JACK NADIN

First published in 2008 by Tempus Publishing

Reprinted in 2010 by
The History Press
The Mill, Brimscombe Port,
Stroud, Gloucestershire, GL5 2QG
www.thehistorypress.co.uk

Reprinted 2011, 2014

© Jack Nadin, 2011

The right of Jack Nadin to be identified as the Author
of this work has been asserted in accordance with the
Copyrights, Designs and Patents Act 1988.

All rights reserved. No part of this book may be reprinted
or reproduced or utilised in any form or by any electronic,
mechanical or other means, now known or hereafter invented,
including photocopying and recording, or in any information
storage or retrieval system, without the permission in writing
from the Publishers.

British Library Cataloguing in Publication Data.
A catalogue record for this book is available from the British Library.

ISBN 978 0 7524 4624 0

Typesetting and origination by The History Press Ltd
Printed in Great Britain

Contents

Acknowledgements		6
Foreword		7
Glossary of Terms		9
1	Bank Hall Colliery	11
2	Carr Drift Mine	30
3	Clifton Colliery	40
4	Copy Colliery	48
5	Deerplay Colliery	56
6	Fir Trees Drift Mine	68
7	Hapton Valley Colliery	73
8	Hoddlesden Colliery	112
9	Huncoat Colliery	124
10	Moorfield and Whinney Hill Collieries	163
11	Nabb Colliery	171
12	Red Earth Drift Mine and Pickup Bank	180
13	Scaitcliffe Colliery	185

Acknowledgements

The transcribed recording of *East Lancashire Mining Memories* could not, of course, have been achieved without the help of those who contributed. Without their contribution our knowledge of the social and working side of coalmining in East Lancashire would be much the poorer, I dare say even non-existent. Special thanks go out to the following in no particular order:

Richard Blades of Stacksteads, near Bacup; Gordon Hudson of Burnley; John (Jack) Richard Counsell of Clitheroe; Archer Lee of Padiham, near Burnley; Stephen Spencer of Rawtenstall; George Greaves of Burnley; Robert (Bob) O'Hara of Padiham, near Burnley; Leonard (Len) Bridge of Rawtensall; Steve Hird of Kendal; Bob Woods of Burnley; Joseph Thompson of Trentham, Staffordshire; Bill Walsh of Accrington; Bill Lonsdale of Great Harwood; James (Jimmy) Spencer of Chipping, near Preston: Bert Holdsworth of Balderstone; David Hargreaves; John Kennedy of County Clare; Ian Walsh of Clayton-le-Moors; Josh Greenwood of Clayton-le-Moors; Stuart Ingham of Belthorn village, near Blackburn; Bob Whittle of Burnley; John Walker of Burnley; Philip Entwistle of Cliviger, near Burnley. Last, but certainly not least, my wife Rita for all her understanding while I was compiling this book.

Foreword

The coal mines of East Lancashire became a major employer from the late 1840s through to the closure of the last deep mine in the area, Hapton Valley Colliery, in 1982. The town of Burnley went on to become the largest cotton manufacturing town in the world, with more looms than population – 100,000 at one time. But it was coal which fuelled the furnaces and the boilers for the steam engines in Burnley and the rest of the country. Without coal Britain would never have become the great manufacturing nation it did. In the early days there was often no practical training given to the young boys who started at the pit. They would spend a few weeks on the surface of the pit, sometimes in the screens taking the stones out of the coal or oiling coal tubs and such like, before being sent underground – but near the shaft bottom usually on the haulage system there. Gradually they would be moved ever nearer the actual coalface until they graduated on to the coalface itself as a collier, although this could take many years.

The coming of nationalisation improved things to a certain extent – but it was never the 'New Jerusalem' that many claimed it would be. At the end of both world wars, the cry went out for 'coal, more coal' as Britain was reduced to its economic knees – and on both occasions the miners responded with greater output from the pits. Their reward came in the 1960s through to the 1990s with mass pit closures, and the skills of generations of miners were lost forever. 'Once a miner, always a miner' it is often claimed, and even today as former miners get together for a pint of beer, conversation of pit days past often crops up. When the pits were running it was said that more coal was moved in the local pub than a week's output at the pit!

I would like to take this opportunity to thank all those who took part in contributing to this book. What follows is a unique – and often tragic – account from some of the individuals who worked in the collieries of East Lancashire which gives an insight into the region's mining history for generations to come.

Jack Nadin

Glossary of Terms

Backening – Shovelling and passing dirt or coal back to the person behind you, which is then loaded onto the conveyor belt or stowed in the gob.

Bait – Or 'bate'. Packed sandwiches eaten at break-time, or 'bait time' when underground.

Banksman – The person at the top of the pit shaft in charge of that general area, and also the person who rings the winder to lower and raise the cages.

Bing – Dirt in a coal seam or an inferior coal band, which often was cut out and left behind.

Deputy – An official underground in a coal mine, next up in class to an under-manager.

Downcast – The shaft where fresh air was drawn into the workings of the mine.

Dowty – Trade name. A hydraulic prop capable of being pumped up.

Drawing – Pushing and taking coal (or dirt) in wagons from the coal face to nearer the pit bottom.

Fireman – The person under the deputy in rank who is in charge of a particular district underground.

Gate – The tunnel at either end of a coalface – these are also called the maingate – where the fresh air reaches the coalface and the tailgate, and the stale air is taken away back to the shaft.

Ginney – A small mineral railway operated by an endless chain, which could be used on the surface or underground.

Gob – The waste area where the coal has been extracted, a dangerous place to be.

Headgear – The steel (or wooden) structure over the pit shaft with the winding wheels, which is used for lowering and raising men, materials and coal from the pit.

Keps – Steel bars which can be withdrawn, on which the cage rests while at the top of the shaft.

Onsetter – The person in charge at the bottom of the shaft taking out empty tubs and putting full ones in the cages. He also communicates with the winder at the pit top by a system of bells.

Shotfirer – A person legally entitled, having passed exams, to fire off explosives underground.

Sump – A collection point below the bottom of the shaft where water is stored prior to being pumped to the surface.

Sylvester – A pulling device used for extracting props. In some parts of Lancashire these are called a 'Dog and Chain'.

Tram – A small, four-wheeled bogie used to get the men nearer the coalface, sometimes propelled along by kicking at the sleepers under the rails.

Upcast – The shaft where the air is extracted out of the mine, mainly by a fan causing fresh air to be drawn down other shafts and ventilating the workings.

Windy Pick – A term used in the Rossendale coal mines for the pneumatic guns similar to those used on digging up road works on the surface, but smaller.

I

Bank Hall Colliery

Bank Hall Colliery at Burnley is the largest of the coal mines to be featured in this publication. The first sod of this new colliery to be sunk by the Exors of John Hargreaves was cut by Elijah Helm on 6 April 1865. The completion of the sinking of the two pit shafts was celebrated by a 'party' whereby 200 men sat down for a meal at the Bull Hotel, the Old Sparrow Hawk and the Talbot Inn at Burnley. A third shaft had been sunk by the early 1890s, and the only seam worked at this time was the Arley Mine at a depth of 230 yards, the best seam of coal ever worked in the Burnley area, which averaged in thickness 4ft of clean coal. In 1913/14 work was begun on sinking the No.4 shaft at Bank Hall Colliery, which at a depth of 1,506ft was the deepest mine shaft to ever work in the Burnley Coalfield – this was completed in 1917. In the 1950s the colliery undertook reorganisation under the NCB. This was an enormous undertaking and cost over £1 million, a huge amount at that time. New battery locomotives were also installed at the No.4 shaft bottom, and a new training centre was opened in June 1953.

The method of working, however, took the coalfaces further and further away from the shaft bottom and increased underground transport costs – by the year ending March 1970 production was down to 330,497 tons, or 39.3 tons per man, per shift. Frequent ignitions on the coalfaces also added to the fact that the colliery was coming to the end of its profitable working life. In the week ending 17 April 1971, the 575 miners employed at the Bank Hall Colliery received their notice and the pit was closed down save for a number of men employed on salvage work. Most of the men took the redundancy money and accepted the fact that mining was ended, although a few did take the opportunity to be transferred to other collieries in the Midlands and other places. Soon after the pit was closed down the site was cleared and redeveloped into an open space, and it is now a place of leisure rather than of toil.

Bank Hall Colliery, Burnley, was the town's largest pit. This scene, depicting the view towards the No.4 shaft, would have been familiar to all the men at Bank Hall Pit. (Jack Nadin)

The purpose of any coal mine is of course to produce coal. The colliers and underground workers are all there to achieve this goal, but there was also an army of 'behind the scenes' workers employed to see through the numerous other tasks needed to keep up the day-to-day running of any pit – they were often classed as the unseen workers at the pit. I am talking about those employed on the surface of a colliery, including the administration staff, the office workers, even the baths attendants and canteen staff. The fitters and the welders employed in constructing new pieces of mining equipment for use underground and on the surface are hardly ever given credit for the work they performed. If a particular item could not be manufactured at their own colliery, then the task would be forwarded to Bank Hall Colliery, whose large engineering works and other workshops would rise to the challenge. The design and technical drawing of these and other projects were undertaken by an elite group of men who were rarely given credit for their very important work.

Bob Whittle
Age at interview   *86*
Years in mining   *33*
Collieries worked at   *Bank Hall, Burnley*

Bob Whittle was one of these men, who spent a lifetime quietly, but proficiently, going about his business in the technical drawing department at Bank Hall Pit. Bob was able to recall with great detail some of the surface features of the old Bank Hall Colliery before it was drastically changed forever during the surface and underground reorganisation of the early 1950s. Here, Bob tells his story in his own words.

I WAS BORN on 15 January 1920, at what is now part of the Robinson Street Conservative Club at No.25 Robinson Street, Burnley. My father was Yates Whittle and my mother Eliza Jane. There was also three brothers of mine, Enoch, John Thomas and Lawrence. I was educated at Able Street School. One of my teachers there was Sam Hannah, the famed Burnley filmmaker. My father was a tackler in the local cotton mills, as were my three brothers. But the mills never inspired me, and after doing a few jobs around town for a couple of years after leaving school, my Uncle Bob managed to get me a job at the Bank Hall Pit on the engineering side of things. I first started off in the wagon shop at Bank Hall. The pits were still under Hargreaves Collieries then of course, and they repaired their own wagons. The main wooden spars on the wagons then measured 12in by 6in and 18ft long [and] made of oak. Timber came to the works by road, just rough cut, almost still trees if you like, and we would have to cut it up to length and width in the workshops. You can imagine the weight of these oak timbers, and we had to get them upstairs in the workshop to cut them down. There was also a big smithy shop at Bank Hall, which had four hearths at the top and a dozen more at the bottom end. I remember these because that Saturday when war was declared I was working all weekend and I had to black all the windows out up at the top of the smithy's shops.

To one side of the wagon shop was the place where they repaired the colliery steam locos – they all had different names. The ones I remember were *Hornet*, *Wasp* and *Bee*: the hives of industry. After a while I went to the joiners' shop. A bloke called Billy Wilkinson was the foreman then, and we would do all the joinery work for the area, for all the different pits in the area. I can also recall some of the surface ginnies that ran at Bank Hall Colliery. One went from the pit towards the bottom end of Tennis Street, and then underground through a tunnel under Colne Road to emerge near Stafford Street, then over the Leeds & Liverpool Canal, then over the 'Ash Pad' on Monk Hall Street onto the main line near the old Bank Top railway station, now the central station. The coal was distributed from here all over the country, and a branch of the ginney went over to the staith on Oswald Street near the gasworks there. A ginney also came from the Clifton Colliery to this place as well. Another ginney at Bank Hall Pit came from the old Rowley Colliery and the Bee Hole Colliery near the bottom of Brunshaw.

Bob Whittle relaxes at his home in Burnley in November 2006 following our interview about his days in the drawing office at Bank Hall Colliery, Burnley. (Jack Nadin)

There was also a wheelwrights' shop at the Bank Hall Pit under the charge of a fellow called Frank Eustace, and he had all the different patterns for every sized wheel you could think of. He would make the spokes and the rims and then place a red-hot iron outer rim over the assembled wheel which would then compress everything into place as it cooled. Practically all the timber here was creosoted in a big tank, any offcuts made right good firewood, to get the fire going at home.

This work was all good schooling for me, with my interest in anything mechanical, and it spurred me on to go to night school at Burnley College on Ormerod Road, and I managed after sixteen years to get two Higher Nationals and City and Guilds in, among other things, carpentry, mechanical engineering and building. Do you know, I have all those certificates somewhere about the house, and no one ever asked to see them if I went for another job, they just took me at my word.

I was asked if I would like to go into the drawing office after this, and I agreed. The chief draughtsman then was a fellow called Dai Price. He was originally Welsh, I think, and liked his whisky, but he came from Wigan every day, and went back every night until he retired, and then Les Parnell took over. I worked in the drawing office on many projects, such as repairing the old coal washers at Bank Hall, and I was also involved with the construction of the bridge in Thompson Park that went over to the old Bank Hall Miners' Social Club. Another unusual feature I can remember at Bank Hall Pit was a little cottage-type house near where the memorial to the mining industry in Burnley is today – I cannot remember who lived there, but I can remember the house. This was demolished during the reorganisation of the pit in the early 1950s.

On some occasions I would have to go out to other pits in the area, such as Hill Top at Bacup and Scaitcliffe Pit at Accrington. We put a bunker up at Deerplay Colliery one time. When Hapton Valley pit flooded there was an urgent need to get hold of a submersible pump to stop the pit from total flooding; this was a job I was put on. Within a few hours I was able to get a pump dispatched to the colliery, perhaps saving the pit from flooding to the roof, or worse still, closure. Eventually the drawing offices were closed down, and they wanted me to go to Walkden, near Manchester, and I did go over to see what it was like. But you see, here in Burnley, in twenty minutes you can be out in the open countryside – but not there, so I did not go. I took the redundancy money in 1969 and got a number of fresh jobs at various places. For instance, I went to Lucas Works at Hapton, where they made parts for the RB211. Then I went to a place at Brierfield where they made bottle tops, but this went bust. I decided then that I would go into teaching, so I went back to college and got a job teaching at Norden Secondary Modern School at Rishton near Blackburn, which is where I retired from in 1985. Today I spend my time doing odd jobs around the house and working on the computer.

John Walker
Age at interview   *80*
Years in mining   *41*
Collieries worked at   *Huncoat Colliery, Huncoat near Accrington,*
                             Bank Hall Colliery Burnley

John Walker was rarely seen by the 'ordinary' miners at the pits where he worked; he was another one of those 'unseen workers' who maintained all the safety aspects associated with coalmining – a job carried out at the weekends or during holiday periods. In 1956 John became the 'Test Engineer' at Bank Hall Colliery, a job which, although 'unseen', was essential to the day-to-day safe running of the area's coal mines. It was John and his companions who tested the winding ropes on the cages, inspected the pit shafts, the cage couplings, the steam winders, the headgear pulleys and looked for any cracks in vital mining equipment.

Although John Walker's family are not local to Burnley, they were connected with the local coalmining industry from the early 1940s. John's father, John Henry Walker, was born in Liversedge between Huddersfield and Leeds in 1903. Although he started off with a grammar school education, John Henry had to leave school early to supplement his father's low wage as a leather tanner, and at the age of sixteen he started at the Diamond Pit, about a quarter of a mile from his home. In 1942 he was appointed as the under manager at Bank Hall Colliery at Burnley when the pit was still worked under private enterprise by Hargreaves Collieries, a position he held until 1955. He was then appointed manager of the colliery for a period of eighteen months before becoming the Area Safety Engineer – a job he had until his retirement in 1968. His son John also became involved in the local coal industry as an engineer, work which included all the safety maintenance.

I WAS BORN at Robertown in the West Riding of Yorkshire on 9 March 1926, and I was educated at the local C of E schools in the village, both at the

John Walker at his home in Burnley on 3 December 2006. (Jack Nadin)

infant and junior. I first went down the pit when I was aged just six years old, at Coates Pit, when my father took me to the pit on the crossbar of his bicycle. When we got underground we sledged in on a little bogie to the coalface, pegging away at the sleepers. When we got to the coalface there were two miners in trousers and football shirts hacking away at the coal with pick and shovels – they had no helmets, just a flat cap, and their only means of light was from a flame safety lamp of the Davy type. The colliers helped me get a piece of coal for myself – I had this for years afterwards until it ended its days on the coal fire at home. I should not have been down the pit of course, this was illegal, but my dad was the manager at the pit. In 1939 I took on a two-year course at Dewsbury Technical College in engineering, maths and mechanics – even though I had failed my entrance to the grammar school. I thought a need to better myself through further education; this was from 1939 when the war was declared, until 1941. I recall that I could not start at the college on my first day because the air-raid shelters had not been completed. In August 1941 I started work at Dawson Brothers, a Gomersal Manufacturing firm which made automatic washers and bottling machines for the brewery and dairy industries. I was fifteen years old then. Twelve months later my father had got the under manager's job at Bank Hall Colliery at Burnley. He actually was on a three-month trial first of all, because Hargreaves Collieries, as it was then, had up to this time never employed a 'foreigner' in such a high position. He passed his trial, even though he was from Yorkshire, and all that remained then was to bring the family over to Burnley. Dad could not get a day off work, it was wartime and all the house removal was left up to Mother and the children. We had a lovely semi-detached house back in West Yorkshire with open views at the back and a cricket field in front, we could see for miles – but then we had to come to Burnley – what a change!

We arrived in the back of Wesley Clegg's removal van on a wet, dreary and drizzly November day with all the furniture and fittings. Coming down Queens Park Road the pit at Bank Hall came into view; my first impression was that we were moving from heaven to hell. Burnley through a smog-filled atmosphere and a wartime blackout was not the most inviting of places. The place they had got for us to live was on Browhead Road, which was across from the screens at the pit, just a broken-down corrugated building really with missing windows. When the screens were running there was dust everywhere and in the wet this turned to a runny slime. Dad was waiting for us when we arrived at about 5.30 at night, then we had to start moving all the furniture into the house. Meanwhile Mother was back in West Yorkshire with little Hillary aged three years and Dorothy aged eight years old. After she had tidied up at the old house she had to get a bus for about three miles

The retirement of Bob Brooks from Bank Hall Colliery in around 1959. Left to right: John Henry Walker, Bob Brooks, William Rawstron. (John Walker)

to Cooper Bridge and then catch the train with all her hand luggage to Manchester Road station at Burnley where she arrived at 8.45 at night. Dad caught the bus up to Manchester Road to meet her. When they came back we had to sort all the bedding out and get the younger ones to bed for a well-earned sleep.

Myself, I had every ambition of becoming a draughtsman, but my dad found me a job with Hargreaves Collieries – not at Bank Hall though, which was just a stone's throw away from where we lived, but Huncoat Colliery near Accrington. 'It would never have worked out,' my dad told me, 'having you working at Bank Hall, because I might have been accused of favouritism towards you.' So every workday I had to walk through Bank Hall pit yard just across the road, and then walk across the iron bridge at Station Approach to catch the train to Huncoat Colliery. Because this was wartime, there were no signs on the railway stations to tell you which station you were at, I had to count the stops: Burnley Barracks, Rosegrove, Hapton and then Huncoat. I was set on at Huncoat as an apprentice mechanic, and in 1947 I passed my exams and qualified. Huncoat was a good pit to work at, it was rather like Bank Hall pit was to Burnley, and it was the central pit for the Accrington area. There were extensive sidings at the pit, where four locos – *Raven*, *Kestrel*, *Linnet* and *Lark* – were all put to work shunting the coal wagons about. In later years they installed Hunslet locos underground at the pit. The colliery was also pretty well self-sufficient as far as workshops were concerned; there were saw mills, joiners' shops, blacksmiths, electricians' shops, tub repair works and mechanics' shops. There was no canteen though; we had to go either to one of the pie shops in the village or to the fish and chip shop. This was always the apprentice's job, to go to the shops. Some of the men though used to bring their own bacon and eggs, and these would be cooked on a little oven by one of the drillers in the mechanics shop.

Because this was still wartime you were under the Essential Works Order: you had to be punctual for work every day or you could be prosecuted, and in fact a couple of lads were taken to Accrington Magistrates Court and were fined. The Essential Works Order also banned any industrial action. There were privileges though in wartime, for instance we got free pit soap, even though we had no baths, and we got tokens for extra bread and cheese.

My job at Huncoat included doing a wide range of maintenance work on machinery such as the winding and fan engines, the coal cutters and the steam locomotives. I was working both underground and on the surface [as] there were no permanent underground mechanics at the pit at this time. The manager at Huncoat was Jack Whittaker; he was also the area manager for the other pits around Accrington, such as Scaitcliffe. His father before him was also manager at Huncoat, and the family used to live at Highbrake House. I was once upset [by] Jack Whittaker during some crack detection on the machinery at Scaitcliffe because we found a defect on the crank pin of the crank arm on the winding engine and had to stop it working until it was fixed.

In May 1956 I got a phone call from Mr Race, the Area Engineer, who offered me the new post of 'Test Engineer' at Bank Hall Colliery at Burnley – this was a new position which had just been created. Prior to this, testing was carried out by Fred Law, a draughtsman from the Area Drawing Office, but he had left the industry to become a lecturer at the Burnley Technical College. I took this job on. The work included all sorts of maintenance of mining equipment, such as the testing of the safety equipment on the winding engines, the suspension gear on the cages in the shafts, including the chain couplings, detaching hooks, rope cappings, and the headgear pulleys themselves. Later I became responsible for ordering all the winding and haulage ropes for the No.4 Area of the National Coal Board. Rather like my father before me when the pits were under Hargreaves Collieries, I was first of all put on a 'trial period' of three months, after which the job became permanent – the pits at this time though were of course nationalised. Because this was a new position a new department was built especially for us where the central workshops were at Bank Hall. As time went on all the cage suspension gear for all the other local pits was tested at Bank Hall. I had a young lad with me who had been injured underground and put with me on lighter duties – but the job could be quite physical at times. Every three months at each pit in turn we would have to test the 'slow banker' on the winding engines, an automatic device which stopped the cages being overwound into the headgear, or the descending cage plunging into the sump or timber balks at the pit bottom. Also every six months 6ft of the winding engine rope, or shaft rope, would be cut off, and this too would have to be tested for cracks and wear. The ropes would have to be tested to a British standard, and would include bending and stretching so many samples of individual strands of the rope to destruction. A further safety aspect was that every winding rope at every pit would have to be changed every three-and-a half-years – whether they needed it or not.

I might add that I was also captain of Bank Hall's cricket team for about six or seven years about this time. We would play our home games on Bank Hall Social Club grounds near Thompson Park – just across the canal from Bank Hall.

I will tell you who was an apprentice of mine for a time, it was Bob Elliot, drummer for 'The Hollies' pop group, if you remember them. Bob kept turning up late every Monday morning because of the late gigs with the band. He was told by Bill Walton, who was the workshop manager, to make better timekeeping if he wanted to earn good money. Well he did earn good money later but not at the pit!

When the Hapton Valley Colliery disaster happened on 22 March 1962 I had been asked by Stanley Holland, the unit engineer, to go to the pit to examine a haulage rope. We had booked a van that day to take us to Hapton Valley. We started work at 8.30 a.m. but the van had a puncture. This was repaired, and we got into the pit yard a little late to be greeted by Jack Frankland, the surface foreman at Hapton Valley, who told us there was a problem on one of the districts with the telephones, and that they could not get any messages out. As he spoke, we both looked across to the upcast shaft and saw smoke coming out of the fan drift top – we both knew then that there was something seriously wrong. Someone then came out of the telephone room next to the lamp room and told us to get as many stretchers as possible and take them across to the top of the surface drift. The news of the explosion was now spreading and soon there were ambulances, police, the fire brigade and other emergency services turning up, and later many anxious relatives – all we could do was man the telephones and try and keep the surface teams informed. My wife knew that I was at Hapton Valley that day, but there was no way of contacting her as all outside phones were blocked by the emergency services trying to get as much information as they could. The lads from the Boothstown Rescue Station were on the scene within minutes, because they had been practising at Huncoat Pit, which was not that far away. The No.2 Boothstown Rescue Team arrived within about twenty minutes all the way from Manchester.

By the early afternoon I was able to get away having done all I could, and I managed to get a lift on a wagon going back to Bank Hall for some equipment. What I did notice on the way back was that all the traffic lights were on red and the junctions managed by policemen, so that the emergency vehicles, including NCB wagons, could have priority. That was the quickest journey I have ever done from Hapton Valley back to Bank Hall – I was then able to inform my wife. It is quite poignant now looking back nearly fifty years to think that those lads going down the pit that day, they thought that it was going to be just another ordinary day at the pit. Many never again saw the light of day; even now it is very touching and emotional to look back on that day.

The coal industry was going through some dramatic changes in the 1960s, and by 1967 there had been so many pit closures that the numbers of areas under the National Coal Board had to be reduced to just a few really, one was Manchester and Burnley, another was St Helens, and then North Wales. Because of this I was transferred to the old Richard Evans offices at Haydock where I became the assistant engineer. This involved moving some of the testing equipment from Bank Hall down to Haydock, and it also meant a lot more travelling for me each day – later we were moved again to Anderton House. It got so that I was doing something like 25,000 miles a year travelling back and forth to Burnley, and to pits in North Wales, St Helens, Wigan, Manchester, even up to Workington in what is now Cumbria. I had got a car when I moved down to Haydock through a loan from the Coal Board, and I was given the opportunity of moving down to that area, but family commitments stopped this. Eventually it all got too much, all that travelling and time away from home, and I finished in 1982. I had done forty-one years, which was enough for anybody – incidentally this was the same year that Hapton Valley closed down, the last deep pit round here. I can never forget the comradeship of the miners. I think the Hapton Valley mining disaster only underlined this solidarity with one another working in the pit; in times of need it was all 'hands to the pumps' as it were. I think that if ever I was in a fix at any time, I would pick a miner to back me up – because I know that they would! Looking back, I enjoyed my time in the coal industry, in spite of the dirt, danger and unsociable hours. I have had the pleasure of working with a variety of colleagues with different backgrounds and interests; newcomers who joined our 'brotherhood' soon discovered that to be accepted you had to have a sense of humour. In most places of work this is preferred but in the pits it is essential. In some situations if you did not laugh you would cry. I have, in the forty-one years I spent in coalmining, shed a tear or two, but the laughs were loud and strong – I feel fortunate to have shared my working life with these friends – 'this happy breed of men'.

∞

James Tattersall ('Jimmy Tat')
Age at interview   69
Years in mining   c.19
Collieries worked at   Salterford Nos 1 and 2, Drift Mines, Red Lees, Burnley and Bank Hall Colliery, Burnley

I met Jimmy at his home in Burnley on 24 November 2006 after he contacted me over the phone about his mining memories. His cosy little flat is full of mining memorabilia and 'Royal' items he has collected over the years. Jimmy was able to recount his days down the local mines through a taped interview which I have transcribed below.

I WAS BORN in Burnley [on] 31 January 1937 on Temple Street which runs behind St Mary's Church on Yorkshire Street, but we moved soon afterwards on to Peter Street on the old Pickup Croft – where Burnley bus station is today. The houses were all lit by gas lighting, and the outside privies down a little alleyway at the back were shared between six or seven families. One privy for six or seven families! It must have been oppressive, but to us it was just an accepted fact of life at that time. My first school was the old Pickup Croft School. I was just about ready for the nursery school though when we moved again, this time to Anne Street off Todmorden Road – I then went to St Mary's School, we all went there. The communities though at this time always appeared to be much closer, there were busybodies I suppose, and the folk coming home drunk at weekends, but they all looked after each other. Mum was a bit of an agony aunt I suppose they would call her today; all the young lasses would come to her for advice if there was a bit of trouble at home or whatever. They all called her 'Aunty Ada'.

I can remember when I was about eleven or twelve years of age, a Mrs Hephrum, she lived at number 61 Anne Street. Well, every Friday Mrs Hephrum used to send me up to Starkie's butchers up Oxford Road with nine pence for some horse meat for the dog. I used to get the three ha'pence

Jimmy Tattersall at his flat in Burnley with a collection of mining memorabilia he has placed in the porch. (Jack Nadin)

change and take it back to her, this went on for ages. It was only years later that I found out that she did not have a dog, it was food for the kids, though they thought they were getting roast beef or whatever. I used to laugh at that!

When it came time to leaving school, there was nowhere else to go except pit or the mills and factories, unless your mum and dad had a few bob and could afford to put you into an apprenticeship. That was very little money anyway, and you had to stay on for six years, I mean my friend Tommy Fallow's mum and dad was all right, so he served his time as a joiner. I got set on at Bank Hall pit, the wages were £2 14s a week, a lot more than what an apprentice was getting. Well, when I had done my training I was put on night shift when I was only sixteen years of age. I had a number of mates at Worsthorne, such as Eddie Kelly and George Hopkins, because we used to go that way to get to Thursden Valley to catch rabbits, and they were working at the Salterford Pit up Red Lees way along the top of Brunshaw. I went to see the manager there, so he told me to leave it with him, and he would see what he could do.

Salterford No.2 Drift Mine near Burnley was one of several pits driven by the National Coal Board at a time of an acute coal shortage in the 1950s. (Jack Nadin)

John Hopkins, one of the top Union men, came round to the house later and told me that Mr Malone had arranged for me to start at Salterford the following Monday. So that Saturday I collected my pit clothes from the lockers at Bank Hall and went over to Salterford Pit; there was no lockers at Salterford, there were racks. So on Monday night I got my checks sorted and was put with a belt-man [called] Bernard Shaw. After about two weeks I was in the baths at Salterford, when this little chap came in and said to me, 'OK, are you a new one?' 'Yes,' I said, because I was a big lad for sixteen years of age. He asked where I had come from and I told him that I had been transferred from Bank Hall. 'Oh,' he said. 'Well my name is Bill Tonge, and I am the safety officer here.' He then asked me how old I was, and I told [him] that I was sixteen. 'What shift are you on?' he then asked. 'I am on nights,' I told him. 'Well you are too young to be employed on nights, don't come in tonight, come in on Monday on days,' he then told me. Well, that will do me, I thought, regular days. Salterford was a good pit, everybody helped everybody else, that's how it was, but it was a wet pit. They were working towards and under the reservoirs at Hurstwood you see. In April 1956 they closed the Salterford No.1 pit down and opened up the new Salterford No.2 pit. It was not that far away, but this was even wetter, actually under the reservoirs. The coal was higher than No.1, but it had what we called 'bing' or a dirt band in the middle. When we had cut the coal we had to throw all this dirt into the gob which made the work a bit harder. Mr Malone was still the manager at Salterford No.2 and Tom Chapman was the under manager, everyone was frightened to death of Tom Chapman. All he had was a belt round his neck, with his oil lamp hung from it and a rail hammer which he used rather like a walking stick. If a belt stopped or anything like that he went mad. One time there was a cutter on the coalface and the weight had come down on it and trapped it. Tom Chapman came up and he started ranting and raving – he were kicking the cutter with his clogs and cursing. Then he set off up the face still shouting and wanting to know why this cutter was not working. Billy Mills was the fireman that day, and Billy got behind some brattice cloth at the top of the face. When Tom Chapman came through Billy hit him with a pit prop in the ribs – it was the only way to calm him down, he was off his head completely. Tom Chapman was a good man, he was a proper pit man, he lived at Mereclough when he was a lad, and his dad worked at Towneley Pit. The father would walk down through Mereclough every day to Towneley Pit where he worked on the pit top. He had a hand missing and a hook in its place, and was employed pulling the tubs out of the cage – they were double-decked cages at Towneley, only small tubs. The tale went that one day he set off for work from Mereclough but never got there. It must have been very strange, because if someone had found him they must have been able to identify him with the hook, but they never found him.

The Salterford No.2 though did not last that long and they closed that down, I think it was because they were short of men at Hapton Valley, Bank Hall and the other pits. I went to Bank Hall, back on night shift again. We were all put into gangs until we got to know the pit bottom, doing different jobs around the pit, helping out with the cutter man, ripping, striking and things like that. Bob Brooks was the under manager – he was a bit like Tom Chapman, he would go mad if anything was not going as it should be. After a while I was put on a face on the Rise, and Charlie Gill was working on it along with a Russian called Alvin. Eventually this became our team, and we were the prop strikers for that face. One time the gob had not fallen for about five or six nights and there was a huge void behind the coalface. What we would do is start in the middle and each work back towards the pack walls at the main gate and tailgate, striking out the props. We would first put up timber props to take the weight and then move the steel props forward. Alvin started to chip the wooden props as he worked his way up the face, so the props would give easier under the weight, and let the gob fall in, we would normally do this last thing. When I started to chip my props, I heard the timbers creak and then the whole lot came in at once. I tried to dive out of the way but the roof fall caught my legs. My light had gone out and as I looked up by the light of the other lamp I could just see the last prop that I was going to chip – thankfully it held, otherwise I would have been killed outright. It was a terrifying experience when that roof came in, the rush of the wind and the noise as all the gob fell in is something I will remember for a long time to come. Charlie Gill came down with some others, and eventually got the stones off me. I was doubled up and one of the men noticed that my hip bones were sticking out of my back. I was in some real pain as the men struggled to get me out into the gate and onto a stretcher. Sammy Marsland, another fireman, came along and asked what had happened. Lol Marriarty, the overman, was also on the scene and it was decided to give me some morphine, which was kept in a concrete box further down the gate. While all this is going on of course I am laid out on a stretcher in all this pain with my hip bones sticking out. Word was sent to the pit top that I was coming out injured and to have an ambulance ready for me, and I was humped on a stretcher onto a bogie. We were at the furthest point in the pit; it was really a long way in. When they got to the end of the rails, they decided that they would have to put me onto the conveyor belt. You imagine the pain I was in each time I went over the conveyor belt rollers on the way out. After what seemed like a lifetime I was raised to the surface and put in an ambulance and taken to the Victoria Hospital at Burnley. The doctor examined me and told me that he thought that my pelvis and hip was broken and sent me off to the Thursby Ward where I was put in the first bed nearest the door. I was examined every hour. A nurse would come and take my blood pressure to make sure that I had no internal bleeding. They did not give me any painkillers though while all this was going on. Later they took me round for an X-ray and the nurse told me to lie on my back. How could I lay on my back with my hip sticking out?! A doctor came along and told me that this was going to hurt and they then lifted me onto the table. I was in terrible pain – I could hear my bones grating together as they lifted me over – but they did get an X-ray in the end. It turned out that I had smashed my pelvis, there was a fracture in my left hip and a fracture in my right hip, which was also dislocated. The doctor also said that he was sure that I also had a fracture at the bottom of my back. I was in bed for six weeks afterwards with my legs strung up on weights before finally getting up on crutches and hobbling about the ward bit by bit. After this I started cleaning my shoes every day and putting them on the bed, trying to hint to the doctors that I was ready for going home – but every day the doctor told me to put them away, and he would tell me when I could go home. All in all I was off work for six months, all I got was injury pay of £9 5s for me and my wife and our daughter Wendy. This was a big drop from my normal wages of around £17. I was still on crutches and Christmas was coming up, so I went up to Bank Hall Pit and asked to see the manager, Mr Watmore, to ask about a job. Watmore suggested that I could work in the lamp room, but I said that I wanted to go back down the pit. If I had gone in the lamp room I would have got the pit top rate, but underground I would have got £14 10s. So he sent me down to work in the mechanics' garage at the bottom of the No.4 shaft. I got some strange looks going down the shaft with my walking sticks – but it was a nice easy number working in the garage just doing a bit of sweeping and tidying up. Sometimes I would have a walk across and talk to Herbert Todd who was working at the bottom of the shaft, or into the cabin and talk to old George Kershaw. I was on this job for about six months before I went further into the pit onto what they called 'Manchester Road' on a haulage engine. 'Manchester Road' was a right steep incline. Herbert Mason, the fireman, found me this job, it was just a case of working the haulage engine pulling the mine cars up full of tackle. I was on this for about two years before I was able to go back to ripping, by this time my compensation case had come up. For all that pain I went through and all the injuries I got they gave me just £1,100, this was in 1963. They were putting the blame down to me, and I was putting the blame down to them – they were saying I should have been using a Sylvester, and I was saying that there were no Sylvesters on the face.

There were some characters at Bank Hall, there was 'John wi' Clogs' (John Halston) and 'Worsthorne Willie'. I was there when Neville Coffey got killed at Bank Hall. We were all coming off on the day shift and as we came to the loco loading point there were some steps down onto the main track. The loco was coming in taking the back shift in, and Neville stood to one side thinking that the train would pass him. But it didn't, it caught him and pushed him under the train. With all the men walking along the tracks over time they formed a small hollow in the ballast, only a couple of inches, and the clearance between the floor on the loco was only about 4in. Well, Neville was dragged under that space, you can imagine what it did to him! He was killed outright of course. Bob Brooks the under manager came and ordered some heavy-duty jacks to be sent down the pit, but the loco was in a really tight place. They could not get the jacks under. They shunted all the mine cars back and tried again – but they just could not lift it. These locos weighed in at about 14 tons you know. Eventually, Bob Brooks did the only thing he could do, he told 'Spanish Joe' the loco driver to reverse the train off him. That was a terrible accident – and of course it happened so quick, it could have been any one of us really. Another time I was working on night shifts

Jimmy Tattersall, back row left, waits to meet Prince Charles on his visit to the 'Weavers' Triangle' in Burnley in 1988. (Jimmy Tattersall)

backening for George Boyle and John Tighe; they would give us half a crown extra each on the Friday, what we called 'pea brass'. Tommy Preston was the overman at the time. A Polish lad used to go in before us, at about half past nine, to get everything ready for us to do the ripping, but one night Tommy Preston stopped us going up. It appeared that all the rock had come down on this lad and he was buried under tons of stone. George Boyle and John Tighe started digging down to try and find this lad. They found his helmet and started to dig round him but it was too late, the poor lad had gone. It was a dangerous place at any time in the pit – when the weight came on the roof some of those steel props would just bend under the pressure and fly out. They would fly like bullets; they would kill anyone who was in the way!

I was still at Bank Hall pit when it closed down in 1971 – it was more or less expected. They were having those ignitions on the face of course, but it was also taking up to two hours to get to the coalface, and then two hours back out. The men were only actually working for about three hours, and it was costing more in getting the coal out of the pit all that way. Some of the Union men argued for a new drift to be driven at Worsthorne, but the Coal Board were not having it. They asked me if I wanted to go to Hapton Valley, but I had made my mind up that I was not going to any other pit, I had done enough. I did a bit with Charlie Gill on the demolition before going on my own into reclaimed timber, with a yard down Whittlefield on Cairo Street. I did this until about 1985, and then got a job with the council training young lads to do groundwork and things like that. We would do things like rebuilding the canal walls and laying out the towpaths. While I was doing this work I met Prince Charles who was doing a tour of the 'Weavers Triangle'. He spoke to me, Prince Charles, he asked me what kind of walling I was doing. So I told him that it was called random walling, rather like dry stone walling only we used a bit of cement. He was alright, and he seemed genuinely interested in what I was saying. After this I went working on the council estates for a while before ending up on the dole, but I was bad with my chest by this time. One day I was signing on the dole, and was basically told to go on the sick. I think this was Thatcher's idea really, because if you were not on the dole there were less unemployed – at least that is the way they looked at it. Some of the things the government at that time got away with! During the Miners' Strike they were bringing the Americans in. One or two of my mates, Alf and his son, were badly beaten during the strike with the police on horses baton-charging them.

I don't do very much nowadays, in fact I can't because of my knees – I have just put my car up for sale because I cannot drive now. I am quite happy here in my little flat. I have my computer, and I do a bit of buying and selling on that, collecting things and so on. I enjoyed pit life though, they were all good lads who worked with you: we all looked after one another – all good lads, yeah.

2

Carr Drift Mine

The Carr Drift Mine was opened up about 1953, a private pit worked under the style of 'The Carr Drift Colliery Company', whose proprietor was Jimmy Cropper who also worked the Carr Farm close by. The coal seam, the Lower Mountain Mine, was worked from where it outcropped at the surface in Shepherd Clough at the top of Dean Lane near Water village, Rossendale. Because it was worked from the outcrop, coal was produced immediately, cutting down on any development costs – the height of the coal was 3ft. The coal was tipped at the surface into coal bunkers and taken away by lorries. Because there was little cover at the pit, it was not uncommon for the workings to break through at surface. On one occasion a neighbouring farmer's cow 'dropped in' the workings. The farmer and Jimmy Cropper spent half an hour arguing over whose fault it was and the cost of the beast and finally Jimmy Cropper said, 'How much furt cow then?' A deal was struck and the dead cow was cut up for meat. The Carr Drift employed twenty-nine men at its peak and was able to produce 2½cwt per man per shift. The pit was closed down about 1969.

∽

James (Jim) Spencer
Age at interview   78
Years in mining   27
Collieries worked at   Nabb Colliery, Water, Rossendale, Carr
                             Drift, Rossendale

The Carr Drift Mine was only worked for about sixteen years between 1953 and 1969 – this is of course is nearly forty years ago, and I never expected to get any more information on the pit. However, Stephen Spencer's (see Nabb Colliery) younger

Carr Drift miners at the entrance to one of the drifts near Carr Farm on what appears to have been a particularly wintry day in the late 1950s. Back row, left to right: Jack Broxton, Brian Seddon, Alan Rawstron, Stanley Binns. Front row, left to right: Jimmy Cropper (kneeling), Dick Cropper. (Stephen Spencer)

brother, James, was able to give me a little more information. Like his brother Stephen, James started his mining career at the Nabb Pit at Water in Rossendale.

I WAS BORN on 5 July 1928 on Burnley Road, Crawshawbooth, the third son of Stephen and Catherine. I left school at the age of fourteen years, and began work the following Monday at the Nabb Colliery at Water in Rossendale, just like my brothers before me. I was up and out of bed at 6.15 a.m. in order to get to work and be underground by 7.30 a.m. On arriving at Water village in my father's car, we would park up and walk the mile or so to the pit top. Here I was given a locker in the pithead baths before going underground. Then I was given five candles to light my way around in the pit, and a tram, which you kneeled on to travel into the pit. I was put with a man named George Cutting – he came from Bacup. I was put on drawing at first, taking the full tubs of coal from the colliers out to the haulage engine and taking empty tubs back to be filled. Pit clothes consisted of anything we could lay our hands on. When I was drawing, it would be an old football jersey and shorts.

The Carr Drift worked a number of inclined tunnels. The last of the Carr Drift mines was at Doals Farm, Bacup. This photograph shows the last drift. From left to right: George Heys, James Spencer and Tommy Helm. (James Spencer)

I can still remember most of the lads that worked at Nabb Pit while I was there: Stanley Binns, George Baldwin, Billy Baldwin, Sam Fletcher, Bobby Parkinson, Tommy and Louis Parkinson, Jack Glenholmes, Billy Corless, John Barker, Stanley and Jack Broxton and so on – some of these came to the Carr Drift when Nabb was closed down. There was not much cover at Nabb Pit, and I remember going into the pit one day, and when I looked at some old workings I could see daylight. The outcroppers at Deerplay had broken through into the old workings. It was not unusual for men just to go up into the cloughs and valley sides and help themselves to the coal where it became exposed and bring it away in carts and wagons. The following morning when we went into the workings up the No.1 District, it was all flooded. The water pumps had broken down where they were outcropping, and all the water had come in from where I had seen the daylight.

There was always a bit of humour at Nabb; they would send the new ones to go and get the 'top lifters' or a 'guttering peg'. There were no such things of course, and many a time the newcomers would just walk around scratching their heads trying to work out what it was.

When Nabb Pit closed down, I was not able to get work in another Coal Board pit, so I went to work in a private mine belonging to Jimmy Cropper, called the Carr Drift Mine. I started here in 1953: Jimmy Cropper was working a seam of coal left by the National Coal Board. The coal was taken away though by the National Coal Board, it was ideal coking coal. We started a few different drifts, one at Carr Farm, one at Clough Head Farm, one at Turnhill Farm and one at Doals Farm, Bacup. This was the last drift they drove. The first of the Carr Drift mines was already there, it was an old stone-arched entrance into the old workings near Carr Farm – that is where the pits were started off. The other drifts at Turnhill and Clough Bottom were started off where the coal outcropped, but at Doals Drift at Bacup we had to drive the drift down from the surface to the old mine workings. The supports we used were wooden props and wooden lintels. The tubs came from the Hapton Valley pit near Burnley, but they were too big height-wise for the seam we were working so I had to cut them down. The haulage engine we used was an old tractor which I converted to run an endless wire rope. I could weld, so I was able to make a lot of the equipment needed for the pit. At the Doals Mine, for instance, I was able to convert a Bristol Crawler tractor as a haulage engine – this was on the surface, but it could be driven by someone underground using cables and pulleys.

I got to the pit by car, and went home dirty; there were no showers at the pits. The pits were in the main dry underground, but a bit loose. Practically all the Croppers worked at the pit, including Richard Cropper, brother Stanley who worked underground, sons Stanley, Morris and Jimmy worked on the surface of the pits. There was also Tommy Helm, Herbert Fitton, Stanley Binns, Johnny Simpson, John Ambler and Stanley and Jack Broxton. To get the coal down, we used compressed air picks which we got from Huwoods at Gateshead in Team Valley. The compressor was from Atlas Compo. The coal came out of the mine in sets of six tubs at a time, hauled up the drift on a wire rope haulage system. These were then uncoupled and then tipped into a large hopper which the National Coal Board wagons backed under. The wagons were loaded by sliding a trapdoor open. Although these were private mines, we were still bound by the rules, and the Mines Inspector would come around every now and again to inspect the workings. The height of the seam at Carr Drift was about 3ft, but by the time we had got to the Doals Mine it was nearer 4ft high. The last of the Carr drifts, the one at Doals Farm, was closed down in 1969. After this, I got work at the Michelin factory at Burnley for eighteen months, then I left there and got work driving lorries for a firm nearer home. I retired from there after working for the firm for thirty years'.

Leonard Bridge
Age at interview 75
Years in mining 10
Collieries worked at *Grimebridge Colliery, Whitewell Bottoms, Rossendale, Carr Drift Mine, Water, Rossendale and Deerplay Colliery, Bacup*

LEONARD WAS BORN on 16 February 1931 the son of Harvey and Edith Bridge, *née* Hey, the eldest sibling to brother and sister Harvey and Mary. Len began his education at Waterfoot Nursery School before moving to Newchurch School, then back to Waterfoot and the junior school and then to the Whitewell Bottom School. On leaving school at the age of fourteen he started work at the Newchurch Boot Company (the factory is still there today) and was employed for eighteen months in sweeping up. Lads were coming back from the army and resuming their old jobs whilst Len was still sweeping up, so he left. Len went on to get a job with a man who delivered leather to the factories, a one-man band with a coal round as well, and Len helped bag the coal. He was fifteen and a half at this time, and when he was

Leonard Bridge outside his home at Rawtenstall in September 2006.

nearing sixteen, Len said to his boss, 'I am sixteen next week', expecting the man to say, 'Well I will give you a rise,' but he didn't, so he left and started to look for work elsewhere. He was able to get a job at the Grimebridge Colliery, then under the National Coal Board, although the pit was worked previously by the Rossendale Collieries Ltd, which offered a 'Man's Job for Life'. When he told his mother, she said, 'You are not going in pit, there are too many rough ones in the pit'. His dad did not bother though, because he knew a lot of the pit lads, who used to live near them, and used to go to the White Horse pub close by. Len got the job by going straight up and knocking on Mr Stephen Landless, the manager's, front door on Booth Road, Waterfoot. Mr Landless was manager for all the Rossendale pits, including Grimebridge, Nabb, Old Meadows and the Stacksteads pits. He said to Len, 'If you want to come into pit laddy go up to Whitewell Bottom, and see them in office'. So he did. The Whitewell Bottom offices faced the old Lambert Works, Osborne Mill and were at the end of the chain road down from Grimebridge Colliery. Len said to the foreman, 'I have come for a job'. The foreman said, 'Have you got your cards with you?' before snatching them off Len and throwing them into the safe.

An early view of the shaft top at the Grimebridge Colliery, Water Rossendale. (Stephen Spencer)

The colliers and mine workers at this time were still on the 'Essential Works Order' and those working in the pits could not get out, even to join up for the army. Some lads were even drafted in from outside, they were called Bevin Boys. Len started work at the Whitewell Bottom coal staith. On his first day he went up the tunnel from the Whitewell Bottom staith and on to the chain road. It was a terrible winter's day; the snow was blowing over the moors and had stopped the surface chain haulage working. All the miners from the pit were digging the snow trying to clear it, and Len was roped in as well. He should have finished at three o'clock but did not get home until seven o'clock. His mother was worried sick; she thought he had been in an accident on his first day. When he came into the house, he was not even dirty – he hadn't even seen any coal that day. When they went back the next day it was as if no snow had been moved, it was just as bad as ever.

Len's first job at Whitewell Bottom was emptying the tubs on the pit top onto the screens, which shook and separated the large coal from the small coal. During the second week at Whitewell Bottom the safety officer came round and told Len that he would have to have some training. This involved one day a week at the Bank Hall Colliery and the Wednesday at technical college at Burnley – this lasted six months. After this he was considered trained for underground work and was sent to the 'Top Pit' at Grimebridge, which was nearer to the building named Smallshaw. 'There were no showers at Grimebridge, no canteen, no lockers, no nothing,' said Len, 'and it was bleak up there.' 'We had to go to and from the pit in our pit clothes, and if it was wet walking to work, you might put a raincoat over the top of your pit clothes.'

At the pit he was put with another 'drawer' and given four candles. Grimebridge was a naked-light pit where candles were allowed: you were given fresh candles every day. There was also a 'clay hole' where you helped yourself to clay for sticking the candle to the tubs, and the lad Len was with took him into the pit. The roadways at Grimebridge were around 36in high, the coal seam itself was 20in high, and the roadways were a little higher because of the height of the tubs. Len's first mistake was to stick his candle on the top of the tub; moving into the pit he was catching his back and his head on the roof bars, and when he went through some brattice cloth used to control the air current, his candle was knocked off the tub and into some water. With no light he could only sit there until someone came with a candle. He got little sympathy and was simply told by the drawer, 'You stick under theer, did nobody tell thee?' indicating to the buffer on the tubs. Lesson number one learnt, Len set off again. Having been told to keep straight on, he passed a number of junctions with roadways going off left and right but he kept straight on. At the very end, about 200-300 yards in, when he could go no further, he was met by the lad who was training him. Len was given four full tubs at the landing here, and told to take them out to the chain road. No one had told him where the chain road was, so he went straight back out towards the entrance, bumping into the fireman coming in. 'What thy doing?' asked the fireman. 'Thy turns off back theer to chain road, gerrit took back.' Lesson number two learnt, and this was only his first day!

Len can remember Jimmy Clayton, Billy Clayton's father, working as a collier at Grimebridge Colliery. Billy later opened up the Grimebridge No.3 Pit and Hilltop Colliery. Other lads Len can remember at the pit were Jimmy Seddon and his son Brian, Terry Atkins, Tommy Briggs, Jimmy Bird and Alan Rawstron. The workings at Grimebridge Colliery went right through the hillside to the Broadclough Colliery at Bacup. When the men were driving this tunnel they were paid so much a yard as they advanced – the fireman would mark each day's advance with chalk on the wall of the tunnel. However, the men would rub this out and put another mark further back. When the management found out they screamed 'We should be nearer Todmorden than Bacup with all the money we have paid out for this tunnel'.

At nineteen Len left Grimebridge Colliery in 1950 when the 'Essential Works Orders' finished and did two years in the army. On coming out in 1952, he and his mate Jim Cartwright went for a job at Hilltop Colliery, a new drift mine opened up by the National Coal Board near Bacup, but they were not setting anybody on so he went to Templey's Brickworks. They had a little drift mine connected with the works to get the fireclay, but the one Len worked at got the coal from the Upper Mountain Mine – this was sent away at the surface in lorries. An old collier from Grimebridge Pit got him set on here, but it was all-day work and little money, so Len was only there about three months. After this he went to Jimmy Cropper's Pit [Carr Drift Colliery, Water]. The pit was working a seam a yard thick, and Jimmy Cropper told him that they had a training face. Carr Drift was another naked light mine, where candles could be used, and privately owned by Jimmy Cropper. These private pits always seemed to be run on a shoestring budget – all make-do-and-mend. Tales are told of how an old tractor would be jacked up at the rear end, the tyres taken off and a wire rope attached to the wheel hub to become a haulage engine to pull the tubs out of the drift. Any old winch or steam engine off trawlers would be utilised and adapted to run off compressed air and put to work at the pit. At the Carr Drift, Len was put on getting coal. One of the old colliers said to Jimmy Cropper: 'The coal's getting a bit harder as we are going further in Jimmy, what we want is a windy pick'. Jimmy was not keen on spending money, but he had to get the coal out. So he thought that he would try it out before committing himself to spending well-earned brass. He managed to get hold of a windy pick by ill means or good, but until he was sure it would work he hired an air compressor, rather than buy one.

The air lines [pipes] were laid out in the drift up to the coalface, and the gun was coupled up. Everyone stood around in anticipation by candlelight to see if this new-fangled thing would do the trick. 'Reight,' said collier Jimmy Seddon. 'I'll have first do,' and so he stuck the pick into the coal, pressed the control handle – and blew all the candles out! After this the men had to use carbide lamps, and did they stink when you had to change the carbide. They did eventually get cap lamps though. Another time Jimmy Cropper acquired a large pile of posts sawn lengthwise and used by farmers for fencing to use as prop bars underground. 'Every time a collier tried to set a prop it kept skidding off the rounded bit and there would be wood bark all over the working face,' said Len. The Mines Inspector was not right keen on those bars either, and he said to Jimmy Cropper one day while doing an inspection: 'Nay then Jimmy, I think we're going to have to find summat better than this', and Jimmy had to change them all – well, some of them.

One day while a drawer was coming down an incline at Carr Pit with a tub, he lost control of it as it was going too fast and he had to let the tub go. At the bottom the tub hit a prop and brought all the roof down on top of the tub. Jimmy Cropper was sent for, and he sent one of the lads out of the pit for a hydraulic jack. Jimmy put the jack under the prop and bar and jacked all the roof fall back up to whence it came – and work was then resumed.

The men working at the Carr Drift one day approached Jimmy Cropper and told him that they were all going to join the Union. 'I will have no Communist agitators here,' said Jimmy. 'There will be none,' explained the men. 'If we join the Union we will be entitled to concessionary coal.' It was the NUM that fought to get concessionary coal, so when those lads at Carr Drift joined the Union they got it too.

Len went to the Carr Drift in 1953, and left there to go to the Deerplay Colliery in November 1956. A lot of the lads were going over to Deerplay because it was now a National Coal Board mine, and things were better work-wise.

Len recalled the time when the gob was 'hanging' for ages, always a dangerous sign as it could drop at any time. While Len was on the face someone shouted out 'The rings are shaking!' and all of a sudden the whole of the waste collapsed with one big clap. Props sprung out and some were sent flying onto the coalface, and coal was breaking off as Len threw himself against what he thought was the only safe place, right up against the coal itself. Happily all the men managed to get off the face without any injuries.

The Deerplay Colliery had one longwall face going at this time, and they were told that the pit had twenty years' work left. But then they began to open up another face, which in Len's mind would reduce the life of the pit by half. When he began in coalmining there were 750,000 men working in the coal mines of Great Britain. He started to think about his future – he was just twenty-seven years old at this time and had done nothing but mining. The future of mining was unsure, with pit closures and cutbacks, and so Len decided to come out of the pits, even if it meant a cut in wages. He found a job working with the fire brigade – a job where he was employed for twenty-five years until his retirement in 1984.

Len enjoyed his time working in the local collieries, and was able to finish his memories of those days with a humorous tale concerning pit drawers. He is not quite sure which Rosendale pit this incident happened at, but it did happen. One day a young drawer who had just started at the pit came to a steep inclined roadway with a full tub of coal. The lad had the common sense to put 'locks on' (bars through the wheels) to act as brakes before going on and pushing the tub timidly over the top of the incline. He hung on to the back of the tub as it gathered speed going down the tunnel. The tub started going faster and faster, sparks flew, but still he held on. The other drawers at the bottom realised what was happening and dived for cover. The runaway tub, with its now very scared passenger, hit the bottom of the incline with a crash and scattered the coal all over the place. The other drawers were now in hysterics, laughing at the young lad, who luckily was uninjured. He stood up and dusted himself off, and, somewhat peeved at all the other lads for laughing at him, bawled out, 'I don't know what you are all laughing at, I will be as fast as you lot one day.'

3

Clifton Colliery

Clifton Colliery at Burnley was located between the Leeds & Liverpool Canal, the railway, the motorway and Clifton Street off Westgate. Today the site of Clifton Colliery has been 'beautified' by landfill from the Whittlefield cutting of the nearby motorway and is now a pleasant tree-filled greenery laid out with enjoyable pathways leading to the Leeds & Liverpool Canal and beyond. Clifton Colliery had a credible history, being sunk in 1876 by the Exors of John Hargreaves, later Hargreaves Collieries, and the owner of practically all the Burnley collieries. There were two shafts at the pit and a cupola, or furnace shaft, for the ventilation at the pit. The pumping shaft (where fresh air was drawn into the workings) was downcast in ventilation and was 14ft in diameter. The other shaft, also downcast, was used for raising and lowering the men and materials and coal – this was 11½ft in diameter. The furnace shaft was just that, and never used for men or materials, and was 9½ft in diameter down to a depth of 780ft. There was a Lancashire boiler, which heated the air around it causing it to rise and draw fresh air down the other shafts. It was in effect a chimney. Although it only protruded around 30ft high at the surface it was the 'longest' chimney in Burnley. The unsuspecting would point out the many chimneys in mill-town Burnley saying that so-and-so chimney was the longest – but none were 800ft long! The pit had at least three surface tramways, known locally as 'ginnies', which took the coal to various parts. One went to the nearby paper mill, another to the Oswald Street gasworks, and another went to a coal staithe across the Leeds & Liverpool Canal on Junction Street. By the 1950s, the colliery was employing over 200 men mining both the Arley Mine and the Dandy Mine – but time was running out for the old pit, and it was closed in December 1955. In the last full year of production the Clifton Colliery mined 39,483 tons, when it employed 219 men.

Clifton Colliery soon after closure, with the 'longest chimney in Burnley' on the left.

Richard Blades
Age at interview 82
Years in mining 18
Collieries worked at *Clifton Colliery, Burnley and Hilltop Colliery, Bacup*

Although today it is hard to believe that anyone could have worked over 900ft underground beneath the grassy slopes on the site of the former Clifton Colliery, it is gratifying that one former miner has come forward with his memories of working at Clifton pit. It was at the Clifton Colliery that Richard Blades began work in 1939. Here, Richard takes up his interesting recollections of his time at Clifton Colliery, and also recalls the dangers involved in mining.

I WAS BORN at 22 Ashfield Road, Burnley on 4 June 1924, the second son of Mr and Mrs John William Blades. They went on to have a family of five boys and seven girls, making a total of fourteen with mother and father. The children were Jack and I, Mary, Edith, Elsie, Marion, Dorothy, Alan, Margaret, Jimmy, Kathleen and Kenneth. Eight of these still survive: four boys and four girls. Times were very hard trying to live off the wage of my father who worked at the weaving shed on Ashfield Road. After a while we

Richard Blades at the author's home in Burnley in September 2006. Richard was the first ex-miner to come forward with his mining memories following various pleas in local newspapers. (Jack Nadin)

moved to 83 Blacker Street, the estate off Barden Lane, and then I went to Barden School on Abel Street. The family was still increasing at this time, so we were obliged to move again, this time to 57 Accrington Road. I then went to Sandygate School on Trafalgar Street, Burnley. The family moved again to 125 Sandygate for the same 'family' reasons, and then to Clifton Street. Meanwhile, Sandygate School was closed down and we were sent to Red Lion Street School in Burnley town centre. At the age of fourteen in 1939 I left school altogether and went working at the Burnley Brick & Lime Company near Queens Park in Burnley, working on the machine which made the bricks. I was there for about nine months, but was trying to get set on at Clifton pit. I finally got set on at Clifton Colliery just before my fifteenth birthday in 1937, which was also just before the outbreak of the Second World War.

My first job at Clifton Colliery was oiling and greasing the coal tubs on the pit top, and then [I] went on the belts picking stones out of the coal. Six months later when I was fifteen and a half years old I was sent down the pit on the level running at the pit bottom. It was not long before I was sent in nearer the coalface and started datalling running couplings props and lids to

Clifton Colliery, Burnley, after closure, showing the cupola or furnace shaft at the pit, 'the longest chimney in Burnley'. (John Walker)

the colliers at the face. My next job was ginney tenting, transferring the coal tubs from one chain ginney to another one. This involved taking the full tubs off the chain and then onto the landing plates, swinging them round and attaching them to the other chain ginney. It was the same with the empty tubs, which were then sent back towards the coalface. Then came the day I was sent drawing for the colliers, and I could not believe how hard it was. The coal tubs were large, and when you left the collier with a full tub, if it was downhill you had to put two steel bars through the wheels, to act as a brake. If it was uphill you had a beret with a bag inside full of sawdust to protect your head and [you] had to pull on the rails with all your strength with your hands. Plus you had to endure the agony of catching the scabs on your back and pulling them off because it was only 4ft high from floor to the roof. It was during this period when I was seventeen that I heard the roof fall, and as I went to investigate some more came down and fractured my skull – I was off work for nine weeks. There was no sick pay; you had to rely on your mates having a collection for you. You would have to wait outside the pay office and the colliers and other workers would each throw a few pence into a bucket. It would then be handed over to you when all the men had got their pay. When I was eighteen, I registered for the forces but was unsuccessful as they wanted men down the pit; coal was important for industry during the war. In fact they were making men go down the pit, these were called Bevin Boys, and so I finished up in the Home Guard. It was during this time that my mother died aged thirty-nine in 1941 after her twelfth child, leaving behind twelve of us.

After nine weeks I went back to work, but could not go back to drawing because of my head wounds, so they put me on a compressed-air-powered coal-cutting machine called a Siskol cutter. It was used for headings or driving the tunnels ahead of the coalface. One day we broke through into another heading or some old workings or something and got flooded out and had to run for it, if you can run when it is only 4ft high and sometimes less.

The foreman said if we get that machine out, you can all finish for the day. We did it by going under water every two minutes because the machine was jacked up to the roof. The cage at Clifton Pit was what they called a 'double-decker' and we went down, crouched down about six or eight men to each deck. We got to the coalface from the shaft bottom by going along the chain road about 300 yards, then we turned right and got on a tram, one

A Siskol Cutter in use. This particular photogaph shows the machine cutting out a dirt band in the middle of the coal seam. The ones used by Richard Blades and others in the Burnley Coalfield would have cut out at the bottom of the seam, which would then be fired down by explosives. (Alan Davies)

of those with a flap which you trapped the rope on which took you along. This took you about a mile into the workings, and then you turned right again down a slope. This went on for about a quarter of a mile, and then we turned right again and went another half a mile, and then we had to walk the ginney roads, one long one and then two short ones. You had to push your tram at one point, by kicking at the rail sleepers with your leg. We had what they called a tower lamp, an electric lamp, but we also had an oil lamp hooked into the top of your shirt – but these kept going out, especially when you were tramming in with the 'judder' from kicking. At the end of here you got directions off the fireman as to where you were going to work that day. By the way, the manager at this time was Alf Walton, and the under manager was Harry Smith. There were five firemen at this time – Frank Thwaites, Joe Ferguson, Alfred Carruthers, Walter Robison and Joe Barford. There were also three shotfirers, Horace Veevers and my uncle's twin sons, Fred and John Hancock. While I am at it, the pit top boss at this time was called Roy Large. After a period using the Siskol machine they installed a larger machine called an Anderson Boyd coal cutter, which was full of small picks and was pulled in to the coalface by two steel wires on drums, one on each side. The wire rope was pulled in by being fastened in each corner by two steel rails up to the roof. When the machine had done a full cut it was withdrawn and you had to drill four holes in the coalface, shoving two drills in with your shoulder. Then you extended the pans, a type of conveyor to fill the coal on after the shotfirer had blown it down. As you cleared the coal, you put a 12ft girder up to the roof, resting it on your back while one of your mates put a prop under the middle. Then it was propped at each end. I was on this machine a number of years until they got a longwall cutting machine, which required about ten men to clear the coal off. One day we were working away from the coalface when the conveyor pans stopped, and we found out that there had been an accident lower down the face. The roof had come in on a mate and it had broken his back. He died while they were carrying him out – his name was Walter Hudson. Just a few months later we lost another collier with a roof fall. He had his head crushed. He was nicknamed 'Yorkie Dick' and was due to retire in twelve months.

A couple of years later we were working at the coalface and the gob, the empty space left behind when the coal had been taken out, at the back of the pans had 'held' for about seven days without falling in. We heard a rumble and the captain on the face, a man called William Spargo, shouted 'Run for it lads!' We got off the face with about a minute to spare. When we went back to look you could not have got a mouse in. I shudder at the thought of that

The last tub of coal raised at Clifton Colliery, Burnley. Left to right are: H. Tomlinson, colliery manager; T. Boardley, banksman; A. Robinson, surface foreman; H. Fort, storekeeper; H.F.S. Smith, colliery overman. (Jack Nadin)

day even now – I went out of the pit and got drunk. But we were back the next day opening another coalface up.

It was good comradeship at Clifton Colliery and it was a happy pit. We had a football team, and it was a good one, I once had some trials for them, but I was never good enough to play. They used to practice on the football pitch on the land behind the old Clifton Barracks. We also had darts and dominoes teams and played in the league.

When we first got married we went to live at the wife's mother's, who lived at No.6 Paper Street off Ashfield Road. We were there about twelve months when we moved to my Uncle Bill's house at No.6 Clifton Street – here we had the luxury of having two rooms to ourselves. My wife and [our] two-year-old son and I had moved to 44 Sowclough Road at Stacksteads near

Bacup in 1953, and I had to travel back and forth every day by bus, setting off at five o'clock in the morning to Clifton pit. Finally the water in Clifton pit won the day and they decided to close it down in 1955.

I was transferred to Hilltop Colliery, Bacup, and was expected to start all over again – but no way after all those years. I did about two weeks, and then they said they wanted me to go on nights, so I handed in my notice and started at the Gas Board in 1956. I did thirty-six years in my working life through different jobs, and finally retired in 1987.

Sadly, Richard died before this book was published.

4

Copy Colliery

Copy Colliery was probably the oldest local pit to make it into the nationalisation of the coal industry in 1947. The pit was located besides the Burnley to Todmorden Road at Cliviger behind Copy Cottages, just before 'Windy Brig', the railway bridge over the former Lancashire & Yorkshire Railway near the Crossing of Dean. Inclined surface tunnels or drifts were driven from the outcrop of the Dandy Mine as early as 1830, almost without doubt by Matthew Jobling, whose family were to be connected with coalmining in Cliviger for generations to follow. The presence of coal here had been proved by numerous bell pits in the area from centuries earlier. The colliery at this time, along with the Railway Pit and the Union Pit at Cliviger, were worked under the title of the 'Cliviger Coal & Coke Company'.

Ventilation at this time was induced by a furnace installed at the bottom of the upcast shaft – the hot air drawing fresh air down the other shaft. Another peculiarity at Copy Colliery was that the winding engine wound 'back to back', that is to say that as it raised one cage in one of the shafts, the other cage was lowered down the other shaft.

By the mid-1930s the Arley Mine was worked out at Copy Colliery, and the decision was taken to reopen the Dandy Mine by driving two inclined surface drifts at the pit. The shafts, however, remained in use as secondary access and ventilation until they were filled up in 1958. Coal extraction at this time was by the pillar and stall method, leaving blocks of coal to support the roof, and the coal was brought down by pick and hand-loaded by shovel. In 1945 the colliery employed fifty-two men underground and twenty surface workers. The pit manager at this time was Sam Jobling, and the under manager was W. Collinge. When the coal industry was nationalised in January 1947, the pit was the last of the old Cliviger Coal & Coke Company's pits. Time was running out for the old pit though, and by March 1964 the colliery had been closed down and all the salvage work completed. The pit employed

157 men before closure. In its last full year of production the colliery raised 57,955 tons of saleable coal. The site of Copy Colliery was landscaped by Lancashire County Council in 1981.

∞

Philip Entwistle
Age at interview   71
Years in mining   16
Collieries worked at   Copy Colliery, Cliviger, Bank Hall Colliery, Burnley

Philip Entwistle is Cliviger-born and -bred and still lives in the village where he worked as a miner for sixteen years, thirteen at Copy and three at Bank Hall. Mining was in the blood: his father before him worked at the Union Colliery, Cliviger, when it was still under the Cliviger Coal & Coke Company. I was able to interview Philip at his home in Cliviger in early December 2006. What follows is a transcript of that interview.

Former Copy Pit miner Philip Entwistle with his wife Shelia, whom he married in 1958, at their home in Cliviger. (Jack Nadin)

I WAS BORN up at Rush Hey, the name of the first row of cottages on the right-hand side up the Bacup Road, at Cliviger, on 6 November 1935. The cottages had a little shop in the middle run by my mother and my father, who was also called Philip – he was born in 1896, and also worked at the pit just before the First World War and just after the First World War. He served as a soldier in France for about three or four years in between. My mother, Nora, was a Windermere girl originally, her maiden name was Harrison. When she first came to Burnley she was a cook in the service of Dr Henry James Robinson. His surgery was at 106 Todmorden Road, Burnley, but he lived at a large house called Springfield House on Todmorden Road, between Springfield Road and Parliament Street, which was almost straight across from his surgery. Mum must have been a good cook because the doctor lived to a right grand old age.

Although I was born in Cliviger, and I still live here, I was educated at St Stephen's School and then Abel Street School at Burnley. I left Abel Street School when I was fifteen years old at Christmas 1950. I had got it into my head that I wanted to work at the pit after they had taken us down Bank Hall Pit while we were still at school – there was not much further education then, and for most it was either the mills or the pit. So I put my name down, and got an interview with Mr Jeremiah, who was the main training officer at Bank Hall. He told us all in his broad Welsh accent: 'Remember boys, you will always have a job for life working in the collieries.' I thought many a time later, that little did he know what was going to happen in later years to the coal industry. My mother did not want me to go down the pit, and when I went home and told her she was not very pleased. Dad was listening, and said, 'Well you have made your choice, it is hard work and dangerous work, and if you do not like it after a while pack it in. But whatever you do, you find another job, because if you do not work in this house you do not get fed.'

Anyway, I started at Bank Hall just after Christmas 1950 doing our usual sixteen weeks' training, and then I was sent straight to Copy Pit – this would be about April 1951. My first week's wages were £2 4s 6d plus 3d expenses. My first job at Copy was running tackle, or supplies, down the drift with a bloke named Ernie Nutter. It was a case of getting you used to going down the pit, and to the working conditions underground. I can even remember the pit shafts at Copy, in fact I helped to fill them in while I was doing my training in 1951. They would send us to Copy and other pits in a big wagon, about ten or a dozen of us, all to different pits, just to give us something to do really. When they sent us to Copy we trainees helped to fill the shafts in. They were not big shafts as far as diameter goes, compared to say the No.4 shaft at Bank Hall, they were just meant to do a job and they did it by all accounts. Later they built the pithead baths near to where the two shafts had been at Copy. We used to have to go home from the pit on a bus in our pit muck for two years after I started, and then they built the pithead baths. The baths superintendent and the fellow in charge of first aid at the pit was called Ted Washington. He was spot on – they trained a team up and they won the first aid competition. He later went to Bank Hall and worked with Sister Waggot, the nurse who went down Hapton Valley on the day of the disaster.

We had a good swimming team at Copy, we won the Burnley Works Sports and there were a lot of teams in it then, Bank Hall, Hapton Valley and all them. In fact Bank Hall had three teams, Bank Hall 'A', Bank Hall 'B' and the workshops team. We would train at North Street Baths, the baths would be full of spectators round the sides and on the balconies.

To get to work I would have to walk down from Rush Hey to the crossroads at Bacup Road End and catch a bus – the fare, which was called a 'workman's return', was about 2*d* I seem to remember. It was a bit of a laugh at holiday times when all the ladies in their nice new holiday dresses caught the bus, and all the colliers got on outside the pit. They

Miners at Copy Colliery, Cliviger. From left to right, back row: Albert Howarth, Sam Jobling (manager), brothers Albert Smith and Harold Smith, Len Cryer and Tommy Pounder. Front row: Harry Coalcough, Geoff Greenwood. (Philip Entwistle)

would shuffle about and change seats trying not to brush against us in our pit muck.

A few weeks after I started, they would then let us go down the pit proper, this might be working on a scraper end, or a conveyor belt or something like that. I would have to watch the coal coming off the face on the scraper and made sure that nothing got caught like a large cob of coal or else the coal would be all over the place. This job also involved cleaning up under the scraper or conveyor belt and stopping it or starting it as it was required. After this I progressed to taking the tackle to the colliers at the coalface. The tackle was placed on small bogies and then run downhill, and then it was uphill, and we would then have to push the bogies with our heads and pull with our arms on the rails. The Dip side at Copy was very wet, when I had to work down there as a collier I was wet through every day. In fact it put one or two in hospital. I was glad to get out of that area, the Dip workings was that wet. It was the roadways themselves that were wet, once you started working on the pillar and stalls it was not too bad because the water used to run past us down the roadways. We got the coal down with a pneumatic pick, an air gun as we called it, and then shovelled the coal onto the belts. It was hard work getting the coal down with the picks, it made a right mess of your hands. I have known blokes with big segs on their hands and if they did not bite them off the skin in between used to split, it got that tight. It was not a great distance to go into the pit, about half a mile or three quarters of a mile. We would have to walk it in, but coming out we used to ride on the belts. This was not allowed, but we did it anyway and jump off if we saw a fireman. The seam was about 5ft 3in at best, but it was in two parts, separated by what we called 'bing' or a band of dirt. The bottom coal was about a yard thick, then this band of bing and then the top coal. This was the Dandy Mine, the coal at the shafts was the Arley Mine but this had all gone when I got there. The coal from the Union Pit further down the road, that was also the Arley Mine – this went to the royal palaces. My father told me that, because he worked at the Union Pit before the First World War.

The coal went to the railway sidings above the Union Pit on a ginney track, and it also went to the siding from the Railway Pit not far from where I live. You could see the railway wagons with the names on, Sandringham and all that, this went to the Sandringham Estate in Norfolk. It was marvellous coal, the Arley Coal, the best household coal you could get. The ginney from the Railway Pit was all filled in with rubble from the houses which were called Lower Damfield when they were demolished. The houses at Lower Damfield were stables at one time, but they were converted into houses for the Cornish miners who came up when our men were on strike – 'knobstickers' they used to call them. In fact Lower Damfield was also called 'Knobstick Row' because of this. The Cornish men used to have to carry a stick to protect themselves from the striking miners. I knew one or two of them, one lived in what they called the 'Cellar Hole' up at Rush Hey, there were only two rooms there, and he was a Cornishman and his brother lived at the bottom of the row where I live now. They both worked at the pit. The Cliviger Coal Company owned all these houses where I am now – but the Jobling family lived at The Grange, that large house down the road near Jack Hey Lane. When the pits were nationalised in 1947, that house passed on to the Coal Board, and when the Coal Board sold it on they also sold these houses on as well as where I live now. The Rawstrons lived at The Grange for many years. I think Bill Rawstron was the manager at Bank Hall, and the production manager – they bought The Grange off the Coal Board. The Coal Board sold all the properties off in the late 1960s to Burnley Rural District Council; the estate agent for them was Eric Halsall. Do you remember him? *One Man and His Dog*. One of the lads I can remember at Copy was Jimmy Bennett. Angus Todd was another Copy lad; he lives at Worsthorne now. Howie Hargreaves was fireman at the pit, Albert Smith was the fireman on the back shift [afternoon shift]. The manager at Copy was Sam Jobling, and the under manager was Wilfred Collinge when I was there. The Collinges were well known in Cliviger, one of them worked in the lamp room at Copy, and another was the landlord of the Gordon Lennox pub. The Joblings as well – they were well known in the village. It was their family that started the Cliviger pits off all those years ago. Copy was a grand friendly pit; it had the feel of being a family pit, sons worked besides their fathers and uncles and so on. There were a few of the Grice family worked there, old Harry Grice was there, he was the father, Sidney Grice was a fireman, and young Harry Grice he started and passed his papers for fireman and shotfirer. I think young Harry went to Bank Hall later on, after Copy had shut down. I think I can count myself lucky inasmuch as I never got seriously hurt while I was at the pit. The only time I came near was the time I caught my hand taking a prop off the conveyor and caught it on the roof. Luckily the prop was just a bit short; otherwise it would have taken my hand off completely – it did split all the palm of my hand though. The roof at Copy Pit was what we might call 'mealy' or 'bitty' and one or two lads got buried under it, like Brian Smith, but he was not badly hurt. I remember one time, I had just started work on the back shift – I would only have been about sixteen years old, and a striker, old Johnny Houghton, got trapped, a girder swung round and took the bottom of his

Copy Colliery swimming team. Back row, from left to right: Harold Smith, Albert Smith. Front row: Walter Whittaker, Philip Entwistle, Bernard Kay, Ernest Haigh, Albert Law. (Philip Entwistle)

ear off. It was a low place, and he had just enough room to swing out the way, otherwise it would have taken his head off. He was buried under a fall and it took about a dozen of us to get him out. He was getting on then, old Johnny, and he never came back down the pit after that, they found him a job on the pit top.

Copy had its own railway siding you know, the coal would go straight over the main road on a gantry and into the hopper on the sidings. Years back if the farmers wanted a bit of coal they would take their carts up there and fill them up with coal – just back them under the hopper. The coal was screened and cleaned just before it went into the hoppers – on the pit top just before the gantry they had men picking stone and muck out of the coal. I think we knew that the Copy Pit was going to close months beforehand, pits were closing left, right and centre at that time. The manager just before Copy closed was John Malone, who had been the manager at the Salterford Pit before they closed that down. He told us one time that he got to know that Salterford was closing down after reading it in the local newspaper. He did not know about it until then – and he was the pit manager at the time. When Copy shut I went to Bank Hall Colliery at Burnley for two years, and then I left. I went to Michelin for four years, before packing in there and going back to Bank Hall, and I was there until it closed down in 1971.

I worked for two and a half years after that with the County Council at the depot in Park Road at Cliviger, then about twelve months at Prestige. I finished my working life though at Norweb jointing cables, spending about twenty-five years there. I think I enjoyed my time at the pit, at least when I was younger, but I used to think I must have been barmy, why did I not go as an apprentice plumber or something like that? But I suppose that working at the pit was something I would not have missed out on really – they were a grand set of lads at the pit.

5

Deerplay Colliery

The Deerplay Colliery was situated high on the moors above Burnley and Bacup behind the Deerplay Inn near Weir village – it was reputed to have been the highest pit in Lancashire. The pits around here probably dated from the 1850s, working small drifts at the outcrop of the seam. From around 1879 the pit was worked by the Cliviger Coal Company, but from 1924 the pit is listed under the Deerplay Colliery Company – this same company was still running the pit in 1945 when it employed thirty-one men underground and seven surface workers. The colliery was one of those nationalised by the National Coal Board in 1947, and some improvements took place in later years. The pit was closed down in April 1968. During the last full year of production the colliery raised 94,000 tons of coal, and employed 212 men. Many of the workers were transferred to other local collieries. Today, only the old access road to the old mine tells of the existence of Deerplay Colliery.

John Kennedy
Age at interview   71
Years in mining   20
Collieries worked at   *Deerplay Colliery, Hapton Valley Colliery, Burnley, Agecroft Colliery, Manchester*

I WILL START from the time I got to Bacup in 1946 and I was eleven years old at that time. We had moved from Ireland from a village outside the town of Foxford in County Mayo, where I was brought up. My father worked in England; he first came over as a young lad. As he grew up he could walk almost every inch of England from down south in Devon and Cornwall all the way up to Yorkshire. He worked all the farms around that area, and Lincolnshire [and] Staffordshire. Both he and his father, and I suppose his father before him, used to come over to England and do farming work. During the summer months he worked on the farms, and during the later months he worked in the quarries between Bacup and Rochdale in Lancashire. When we moved to England my father took on this work full-time. He spent the rest of his working life in the quarries – work was hard in the extreme, he would work seven days a week, and he had just two days off every year, Christmas Day and Good Friday. All this hard labour caused him to finish work aged sixty-three through ill health – he only survived another five years, and died aged sixty-eight in 1972.

When we came to Bacup in 1946, the family consisted of my older brother Michael, younger brothers Eddie, Tommy, and the youngest Stephen, born in England, and sister Mary. Unfortunately Stephen died when he was just eight months old. At Bacup we went to St Mary's School. We lived on Plantation Street in the town – know better in Bacup as the 'Irish Back'. Plantation Street was one long street just off Market Street – the only view we had was a wall, a wall maybe 10ft high, holding back a large slagheap. You cannot see the slagheap today, it is all covered in trees. When we were younger there was just a few trees, and we used to play up there most of the time. From this vantage point we could count all the factory chimneys in Bacup: there were thirty-one of them. Besides the cotton factories there was also the slipper works in Bacup, where they made all the shoes – it was a busy old town at that time. They would all finish work about half past four in the evening and the streets would be crowded, really crowded, even the buses had to crawl through the streets and the crowds. To take the workers home there was up to three or four different services coming into Bacup, one to Rawtenstall, one to Haslingden, and one to Accrington, one to Todmorden and one to Burnley and Rochdale – buses were leaving the town every five minutes, every one of them crowded with the workers. The younger ones used to walk up to the centre of town, even though this was the opposite direction to which they wanted to go, those who stayed behind might have to wait up to an hour because all the buses coming from the centre were then full up.

When I came home from school at dinner time I used to go to the Co-op shop on Market Street. Things were tight; there was little money about, and even less food. One day the manager at the Co-op followed me out of the shop and asked me if I wanted a job. 'Sure,' I said. Sure I wanted a job. 'What do you know about Bacup?' asked the manager. 'Not much,' I said. So the manager gave me a list of Bacup streets and told me to go and look for these streets saying, 'We will see what we can do for you'. After a week or so, by asking around, I did find all the streets the manager had set for me. I had to go into the Co-op every day, and the manager would say to me, 'Have you been looking for those streets I gave you?' 'Oh yes,' I would say 'I know where that street is, and that one, but I still have a few to go yet.' Eventually I managed to find all the streets, and I knew where they all were – so I went back to the shop and informed the manager. He said, 'Right, this is what I want you to do. On Wednesdays and Thursdays people come in the shop and they want their groceries delivered, do this and there will be a few bob in it for you.'

I took the job on – it was great, I might get 3*d* at one delivery just for walking half a mile, or a mile, and then to the next one, and it was the same there, another 3*d*. After some time, I was delivering groceries for two other shops, I also had a paper round, and had to deliver the newspapers before they went to work – for this alone I was getting 6*s* a week. I was also getting to know Bacup town. Coming from a little farm in County Mayo, Bacup seemed to me to be a huge different place, but slowly I was getting used to it. The house we lived in at Bacup was a two-roomed house, one bedroom and one room downstairs, the kitchen was a space about a yard square at the bottom of the stairs. Here we had an orange box, and on this was a gas ring where all the cooking was done. There was also a slopstone with a hole in the middle, which passed as a sink, there was only one tap, a cold water tap, and if you turned this on too fast there was water all over the show. At the top of the stairs was a small landing, and my father made this up with a bed for my brother Eddie and myself, my older brother Michael slept downstairs on the sofa, Mary, Tommy and Stephen slept in my mother's room. When my older brother Michael left school at the age of fourteen, he decided that he wanted to go to work back in Ireland. There was work available in Dublin in the building trade, so he went back there for a few years. I left school in 1950. There was plenty of work around at the time in Bacup, but I wanted to be a bricklayer – it would be nice to build a house I thought. So I managed to get an apprenticeship with a property repairer, doing a bit of slating, repairs to roofs, a bit of plastering and bricklaying and so on. It was nice, I enjoyed it. I was only labouring, but I thought with a bit of luck I might pick something up. I only had the basic education, but nevertheless it was decided that I must go to night school. I was not learning anything really there, so I went to the teacher. He was only a young man, who had done his National Service, and asked him if he could give me a hand, as I had not done any of the subjects we were working on. He told me that when I had done my National Service, to come back, and that he would teach me all I needed to know. I used to have to go to the grammar school at Waterfoot for this so-called night school education, on my bike there and back, rain or shine – so the proposition by the teacher put an end to that. I did carry on with the job, the money was not that great, about £2 or £3 a week maybe, and the hours were long. By the time I reached eighteen years of age I was called up for National Service, and with another lad from Bacup we went down to Worcestershire to do our training. We got to come home on a forty-eight-hour pass back to Bacup, and we were there in our uniforms, the first full suit of clothes I had ever worn, I was really proud. When we got back to camp, we were told that we were splitting up, some going one way, others the other way. This split up the friendship I had formed with other lads from the Rossendale Valley. We did meet up again when we came out of the army though.

Eventually I arrived at a camp in Stratford-on-Avon, but I did not get on very well there – the sergeant major took a dislike to me, and I was given something of a hard time – thankfully I was not there very long, because I had put in a posting to Korea. I was posted to another camp, and called into the commanding officer's, and he told me that there had been a change of plan, and that I was not going to Korea, instead I was going to Germany. On the way to the Hook of Holland, I met a gang of lads who had been to Korea. It was a rough crossing and a long time in getting there, many of the lads were seasick. When we landed in Holland we caught a train overland to Germany. Looking out of the window was like watching a film: the war had been over for eight years, and yet the wreckage and destruction in the countryside was still very evident. I could not believe it all. Germany itself was even worse, towns in ruins, the countryside devastated. To cap it all, it was wintertime, I had never been so cold in all my life. One strange thing I noticed was that the snow was very dry, so dry you simply brushed it off yourself – but it was cold, very cold. The camp itself though was pleasant enough, warm and friendly, with gyms and all sorts of things – but you had to be very hungry before you could eat the food. The wages were about 10s a week, or something like that. By the time you had bought your soap, toothpaste and blanco you did not have a lot left over to go out and have a good time. It soon came to the day when we could go home – it was strange, when I was away from home I missed it, but when I got home I wished I was back in Germany. Back in Bacup the family now had a Corporation house up on a place called Fair View, this got a bad reputation in later years, but then it was a lovely place to live. This is when someone suggested that I try and get set on at the pit – I had never thought of working down the pit. I knew there were pits about, I had seen the lads going home all black. So off I went to the dole office and got set on, but for some reason instead of going to Bank Hall Pit in Burnley where all the other trainee miners were sent, I was posted down to Stoke-on-Trent to a Coal Board hostel – it was like an old army camp. There were other lads from Bacup there, which helped a little. The wages were £7 a

A group of Deerplay Colliery workers enjoy a pint on a day out to Blackpool. From left to right: Alan Clawson, Derek Copeman, Derek Marshall, Clifford Pilling, Vincent Chadwick, Mick Clegg, Tommy Moffat, John Kennedy. (John Kennedy)

week, but it cost £4 a week to stay in this hostel. Food consisted of spuds for your breakfast at half past four in the morning, and the same thing for your evening meal at half past four in the evening.

The 'training' consisted of a day underground and a day at the college and we would also go working on the slag heaps. This was worked by a rope haulage, and our job was to tip the tubs up at the end to empty them. It was here that I saw something I had never seen before – way down at the bottom of the slag heap were women, children and old men picking coal from the heap and bagging it up. Some of them were probably selling it, others were taking it home for the fire. If we were not taking care emptying these tubs, there was a danger of hitting these people collecting coal down the bottom. I could only assume that it was desperation that forced these people to risk their lives like that – the need to keep warm in the home. There was a security man who used to go around and chase them off with a big stick. For all the world from our vantage point it looked as if he could well have been a farmer chasing crows off a field – as soon as he was out of sight they all came back again. I finished the training, and I was told about a pit I could go to which was way out in the countryside. There was a bus which would set off about half past five in the morning and it did not come back then until about four in the afternoon, or something like that. It was a typical mining village, all built around the pithead, but wonderful folk. A

Deerplay Colliery

few others and myself were taken out to visit the pit, and the manager came to welcome us. They were short of young men at this pit, and he said to me, 'Any one of them houses there, just pick the one you want and you will be quite welcome to have it'. I did not like saying at the time, but I did not like the place at all, there was nothing around! It was while I was doing my training that I used to go home to Bacup every weekend, and I would look round the local pits there; there was Deerplay Pit of course, Hilltop Pit and Grimebridge and the Old Meadows Pit. Old Meadows and Grimebridge were very low, and used the old hand getting method of getting the coal – to my mind I would have preferred either Hilltop or Deerplay. So, one Saturday I went up to the Deerplay Colliery, where the manager said he would give me a chance. I went up to Deerplay Pit, which was completely different from the Staffordshire pits, being a drift mine of course.

It was all a new experience to me walking into the pit rather than going down in the shaft. On my first day I was taken in and set to work with Jack Heys and Paddy Burns, two lovely lads to work with. We worked on a machine we called by the nickname of 'Jim Crow' which was used for bending the rails, the tub rails, to make them go around the corners. Well, the first day I was a bit stiff and that, and I took my time in the shower and I happened to miss the bus. So Jack Heys took me down the ginney road which took us to Weir village where he lived and he invited me into his home to wait for the next bus. The buses only ran every half hour or perhaps every hour, anyway his wife brought us a big bowl of broth each and a pint pot of tea. I could only think to myself, what a lovely thing for a man who I had never met in my life to be so kind and generous; even now, to this day, I just cannot believe just how kind people can be. It turned out that they knew my father from when he first came to Bacup as a journeyman navvy all those years ago. This carried on for a good time after this, a good pot of tea at the end of the shift and a fine bowl of hot broth – it was a lovely thing to have. But after a time I was getting better at the job, and was out of the pit quicker, and I must have let it slide. I should have kept in touch with those beautiful people. I still saw Jack every day of course at the pit, but his wife was such a lovely person and I am awfully sorry that I did not keep in touch with her.

As I got used to pit work I was placed on different jobs about the mine, working on tackle running and so on. We had these little trams, four wheels upon them which you could pick up and put under your arm and carry along if you were going from tunnel to tunnel. You could put that down on the rails and put three props on it and then put a galvanised sheet on top of that, and then you could put about twenty or thirty props onto the tram. Once you knew how to load them you could put prop lids on, steel bars, chocks or whatever and take them to the coalface. Sometimes we would take the rings on, rings for three shifts to the coalface, and we would have to get down on our hands and knees and shove with your head – just like climbing a ladder, pulling on the rails and shoving with your legs on the sleepers. Most of the time it was uphill, climbing uphill to the coalface. But of course we would be riding out on the trams coming off. Sometimes though we might have to go down the brew, in this case you put two wooden sprags in the back wheels to act as a brake. The only problem with this was if you came onto some wet rails, in which case you would be going like a sledge, then you would really have to work to keep the weight back. By this time we were on contract work, with quite a bit of tackle to move, three lads to each district. The tackle had to be at the coalface before we got our wages, which at the time I think was about £4 a shift – good money for the time, but very hard work. The team was Tommy Moffitt, Vincent Chad, Derek Marshall, Andy Hannegan and myself. By this time I was married and I had two children, so every halfpenny counted. After a while, being the senior lads, Tommy Moffitt and myself were chosen to go face training – I could not wait to get face trained, but the only thing was that your wage would drop to £10 a week, an awful slump in wages compared to what we were getting on the tackle running. With the wife and two kids it was going to be a real challenge for the next six months, but we managed it.

Up to this time you would be trained for just one job on the face, be it coal getting, ripping, backening, pan moving or packing. Some of the lads had just come to Deerplay from the closed-down Copy Pit in the valley below – they were just colliers, they just filled the coal, that is all they were trained for, they had no papers for anything else. These included Sid Mitchell, Arthur Dole, Harry Mason and Mitchell Lord – all grand lads, when Copy shut they came to Deerplay. I started my training with a man called Fred Nuttall, a Burnley lad – grand fellow. However soon after he finished and got a job in Burnley somewhere – funny I never saw the lad again, he was a grand lad. I took well to the training and was put with a man called Peter Ormerod from Bacup – a lovely fellow and he took care of me as if I was his own son. I was learning all the time from him. I was alright shovelling with my right hand, but you had to shovel with both hands down the pit. He would say to me, 'Nay, now, use your left hand not your right'. He kept me safe, he taught me how to do everything – I appreciate now what that man taught me. I got my face training done, this included striking, coaling, [using] the coal cutting machine, and ripping. At first I was on £12 a week which rose to £15, but I was in it for the money, you used to get 1s 6d a day for carrying the powder canisters in, so I did that as well.

Mining is a strange occupation, all would be going well until say the Thursday morning, then the pans would break down, or the machine would be stopped, and with that went your bonus. At Deerplay we were hand-filling for quite a while, each collier had a ratch of coal 8 yards long and you went in there and cleaned it out, you got it off any way you could. The face captain would go through the face every morning and chalk out your coal, called a 'ratch'. You first made a hole for yourself, and then got stuck in, you did not stop then for anything until you got your coal off. Sometimes you could look back into the gob, where all the coal had been got, and see the shiny floor and the shiny roof and not a prop up. Once you had cleared your ratch, you made sure the props were moved forward and set. There were two men on the longwall coal cutter, one man would bring the rope out, which hauled the machine along the face, and attach it to a sprag, a prop set at an angle. The man would then nod his light as a signal to the machine man to start up the cutter, the strain would be taken on the rope and the machine would start to undercut the coal. The rope man would then set sprags under the cut to stop it coming down too soon. The colliers then cleaned up the fine dusty coal from the cutter, called scufftings.

Mining machinery could also be used on the surface of the pit, as in this instance, when John Kennedy used a bucket loader for building the manager's garage at Hapton Valley Pit, near Burnley. (John Kennedy)

It was the job of the men on the afternoon shift to drill the coal above where the cutter had been at work. Alfie Murray was one of them, a Burnley man – a lovely fellow, and I drank with him many a time in Burnley. The night shift fireman would then fire the coal down. The day shift would then come in and fill the coal, and that is how the system worked at Deerplay Pit. The coalface where I was working was coming to an end, and they had just started up a new system with the Anderton Disk Shearers. This was light years in advance of the system we were using previously – the hand-filled system. You were never still after this, it was a lot more work, although the wages went up a little bit. The whole face team were then booked as 'spare men' to be put on any job which needed doing. Peter Ormerod, Rennie Thomas and myself were put on a team doing back-ripping along with a team of Polish men, who were going to put a bunker in. This was in case the conveyor belts stopped for any reason; they could at least then get the coal off the face and into the bunker. We had to work on a scaffold with planks in order to get up and reach the roof. There was a fault, a split in the roof and a huge rock about the size of a single-decker bus was just hanging there. We knew it was dangerous, and we all stuck together, then all of a sudden it came down. This huge rock just dropped from the roof. The first thing to go was the scaffold planks, and they were shattered like matchsticks. Happily no one was hurt, but we were left with the predicament of how to get rid of this big rock, which was now in the way. In the end we drilled and fired it into smaller pieces and sent it out of the pit that way. Next day a man named Ernie Hawk took my place in that team and I was put on another job further into the pit. We were bringing supplies down for the new face, Dowty props, chocks and timbers and the like. The lad I was with this day was Billy Clayton; a great worker, and a marvellous lad. He went on to open his own pit in later years. Pit work was the norm for Billy, he would work just as hard on day wages as he would on contract work – he was that sort of a worker. So I knew I would have to be ready for a busy day that day. We had just loaded the tackle bogie and was ready for setting off, when we heard the air door slam. So we waited a while to see if anyone wanted to come through the doors, when the safety officer at the pit came screaming through the doors shouting to get a stretcher and that there had been a roof fall. We grabbed a stretcher and ran down. When we got there the dust was still swirling about, but we could not see anyone else around. The roof fall was obvious, it was right in front of us, so we started tapping on the pipes to see if we could get an answer, but none came back. It was not a large fall, it just seemed as if the roof had come down and crushed a few ring supports. We did not know what to do, so we started shovelling; we had nowhere to stack it, so we just threw it behind us.

Soon after this more men started turning up, and we were able to form a 'bucket team', still throwing the material behind us. As soon as one man was exhausted another would take his place – that is how it was in mining. Soon there was more muck behind us than in front of us. It was decided that the best option would be to shorten the conveyor belt and send the muck straight out of the pit. Moving all the muck left just a large hole in front of us, so we also had to start setting more supports to cover ourselves. By this time the men had been working for at least two shifts, and the management kept sending down buckets of tea and sandwiches with the tackle lads to keep us going. After a while we could see the tops of the arches, it was all dusty and black with lots of loose stuff still falling in – small roof falls. Tommy Murphy, Jimmy Chad and myself then started to come across the men who had been buried – the first was Frank Vesick and of course he was gone. After a while we got to the second man; he too had died under the rock fall. His name was Eddie Brossie.

Another man, Mario, was very badly injured and had to be dragged out; he never went down the pit again. The only way we could repair the roadway afterwards was to put up some arches and blow concrete behind and into the cavity. The cavity was so large it took well over a month to fill back. It was a day many Deerplay men will never forget. [This accident occurred on 24 June 1964 – Ed.] The bunker was installed in the end.

There was one other accident which comes to mind, and that was on the face. Bob Belshaw and his son David worked at the pit. There was a bad fall on the face and Bob got trapped and killed. They were the three men I remember getting killed – I worked with them all. I worked with the Polish men while I was on the button, and I worked with Bob on the coalface. Bob was a lovely man, he lived in Stacksteads near Bacup – I knew both him and his son very well. We were used to accidents in the pits, the odd broken arm, a trapped limb, scalp wounds, but these times were the very worst. Deerplay Pit was like there was so many brothers, fathers and sons, uncles and cousins, they all worked together like one big family.

I can honestly say that Deerplay Colliery was the first place that I ever enjoyed working at – we were all decent friends, pals that looked after your back without reward or gain, which was never expected or given for that matter. All one, and all for one another – this was mining at its best, a true comradeship, a proud tradition, and one I was pleased to be part of. I once got a photograph from my sister, a picture outside the Deerplay pub, and the newspaper wanted to know what it meant, what it was about. It was the last shift at Deerplay Colliery with the men having a farewell drink. Funny enough, the people passing on the buses (there was a bus stop outside the

Deerplay Inn) used to think 'what was going on at the factory at the end of the road there?' They could hear the fan going even on the bus, and they were wondering what sort of factory it was at the far end of the road. There was two factories further down the road in Weir village, they thought Deerplay Pit was a factory, 'tis laughable. I still look at the picture today and see the people and their names all come rushing back to me. The Deerplay Pit closed down in 1968. I would say I was there about eleven or twelve years maybe. The lads that were on the picture, there was Austin, I can never remember his second name – he was a fitter – George Ormerod, Billy Holt, Keith Whitehead, Keith Thompson, Billy Crowther, Billy Murray, Jim Lord, Billy Stead, Ernie Hawk from Burnley, John Corbett, Jimmy Copeman, Jimmy Dawson, Ralph Pilling, Johnny Mack, Jock Eastham, Joe Corless, Phil Lockland, Rennie Thomas, Tommy Birk, Mal Stott – that man died, he was only a young lad, a Bacup lad – John McNulty, Brian Pilling, Derek Austin. There are so many of them, those are just a few of those who worked there, and I must not forget Paddy Mardy, who I also worked with. It is strange that when I see the picture today, they are all young people inside my head – but an awful lot of them are gone, and yet they were some great people. Lovely people to know, and I am proud to have known them. I went on after Deerplay had closed to Sutherland, to Bank Hall Pit at Burnley, to Hapton Valley Pit at Burnley, and Agecroft Colliery at Manchester. Yet there were still familiar faces from Deerplay all the way through these pits; there was Bobby Hawk from Burnley, I worked with him at Deerplay, with his father as well, Brian Pickles; a Bacup lad, he went on to Hapton Valley and then Agecroft. 'Tis a lovely thing to look back on them after all these years, they were such a grand bunch, a grand group of men. Wherever you go, to look back on men like these it was a lovely time. I remember when I worked up at Bank Hall on the afternoon shift. One of the lads on the face offered to take my place, because he wanted to be a spare man, and I happily accepted. This meant I could be on a regular shift, and that was great for me. The only problem was catching the bus back to Bacup after the shift; I would have to wait up to an hour. I was standing on the bus station one time and a bunch of lads just started coming off the afternoon shift, and they started dogging me with pints of bitter from the Boot Inn. I had been doing this for a while then, so I started to go round with them to the pub. They were all dressed up and clean; me, I never thought about it, my hair was still wet from the pit baths, and I had the make-up around my eyes, the black from the pit dust, like mascara, I must have looked a real state. Anyway, I went in the pub with the lads, but I had to hide away up a corner while they were having a chat and a bit of a laugh. Later on of course the lads moved on to do the town and I was left on my own. Then something strange began to happen. Strangers in the pub would come up to me and say 'So, you work at the pit do you?' and other sorts of conversation like that, and in the end I became accepted as part of the pub company. It made me feel warm on cold, wet nights, and I enjoyed the crack, so much so that I would make my way over when I had the time, all the way from Bacup. As time went on and Bank Hall Pit had finished, I knew all the lads in the pubs, the Boot and the White Lion, it was grand.

When Bank Hall finished I went to the Valley, and I met some grand lads there – but I suppose some of them will be writing about that pit. From Hapton Valley I went to Agecroft Colliery. Here was a lad called Walter Scott and another named Craig Walsh, they were the team mates I worked with there. We worked on the development, we did some good work, and we broke the Lancashire advance record driving a tunnel. We were driving a 20ft by 20ft tunnel, although it was said that we broke every rule in the book. The daft thing was, though, as soon as we had finished all this, they decided to close all that area down because the coal had gone too low. They were used to the coal there being 10 or 11ft thick, and in that place it had dropped to about 6ft high. All our work was for nothing, but that is how they were treating us at that time. All they were interested in was wasting money and closing the pits down – and that is what happened at Agecroft Colliery. When I went back to Ireland in 1991, I was walking miles every day, I could not sit still, and it took me a long time to settle down. I am living now about 100 miles from where I originally came from, and every couple of weeks have a run up there to visit my cousins and old school pals. I like nothing better today than having a gentle stroll in the countryside with my dog; we have lots of open country around where I live. In other words I just take things easy and as they come – what could be nicer than that?

In typical beloved Paddy fashion, John finished off his contribution with an Irish blessing, which went something like this:

May the road rise to meet you
May the wind be always to your back
May the rain fall gently upon your field
And until we meet again,
May the good Lord hold you in the palm of his hand.

6

Fir Trees Drift Mine

Fir Trees Colliery was another of the little drift mines opened up by the National Coal Board, this being the last of three such drifts driven near the outcrop of the Arley Mine around Fence and Higham villages, near Burnley. Work began in driving the drifts at Fir Trees in early 1958, and after extracting almost 160,000 tons of best Arley coal, the pit was closed down on 11 March 1966. Most of the sixty-eight employed there were transferred to Hapton Valley Pit.

○○

Gordon Hudson
Age at interview   62
Years in mining   5
Collieries worked at   Bank Hall Colliery, Fir Trees Drift Mine

Gordon Hudson comes from a family with a long and proud history of coalmining in East Lancashire. His grandfather, Charles Edward Barrowclough, who was born in 1899, started at the old Towneley Colliery near the top of Todmorden Road, Burnley, when it was still under the ownership of Brooks & Pickup. Charles Edward Barrowclough remained at Hapton Valley Pit until he was forced to give up work before the official retiring age through ill health. Gordon Hudson's father Albert, born in 1920, was also in the pit. He started at the Towneley Park Pit on Broad Ing below Towneley Hall at the age of twenty-four before moving on to the Salterford No.1 Pit, and then finishing at Salterford No.2 Pit. Altogether he did thirty-five years at the local mines, twenty-two of those as a collier at the coalface. Apparently the shaft was so shallow at Boggart Brig Pit that you were able to shout down the pit and have a conversation with those at the bottom.

Fir Trees Drift Mine

This photograph gives an indication of just how steep the surface drift was at the Fir Trees Colliery, near Higham village, Burnley. (Jack Nadin)

GORDON HUDSON WAS born on 26 November 1944 at Dall Street in Burnley Wood, the eldest son of Joyce (1924-1996) and Albert Hudson (1920-1994). His brothers were Leslie and Brian. Leslie went on to work at Bank Hall Colliery at the age of fifteen in 1972 and served his time as an electrician along with the compulsory face training. When he left the pit, he went into the Merchant Navy, and then onto the oil rigs before going to Bermuda. He's back in Burnley now but still works as an electrician. Gordon Hudson was educated at the Burnley Wood Infant School, St Stephen's Junior School and Burnley Wood Secondary Modern School. One of his teachers, named Binns, played rugby for Burnley. He recalled happy youthful days of

Gordon Hudson poses beside the mining certificate belonging to his father, September 2006. (Jack Nadin)

long, hot summers and childhood games played in Burnley Wood. He left school at the age of fifteen and got a job at Mullen & Durkin, the building firm on Trafalgar Street, as an apprentice bricklayer. After about a year he asked his grandfather if he could get him a job – which he did more or less the following day. Like most miners at this time he had to do sixteen weeks' training, a week at the Municipal College on Ormerod Road and a week at the Bank Hall Pit alternately. He continued his bricklaying course at the pit after this, on two-day courses each week for two years, and got to and from the pit each day on his motorbike. The first time down Bank Hall No.1 shaft was a frightening experience for Gordon, and for many other trainee miners. The winders at Bank Hall appeared to have a sick sense of humour, and seemed to know when trainees were in the cage: they would lift the cage slowly off the keps, and then lower you for a few feet before dropping the cage like a stone.

Gordon's first week's wage at the pit amounted to something like £5. He later went to the Fir Trees Colliery off Fir Trees Lane near the Pendle-side village of Higham. The main Union man at Fir Trees was Bob Wilson, known as 'Red Bob' who in spite of his 'Communist' ways was always good for a laugh. Another Union man was John Lord – he was also the delegate for

Reedley Pit. The drift at Fir Trees Pit was very steep, too steep for coal to come out on a conveyor, so they had to install a mine car which was hauled up the main drift. The coal seam itself was also very steep, but this lessened as more and more coal was worked. Work for the colliers was very hard here, and they had to use the pillar and stall method of mining using compressed-air guns, the headings being fired down by the shotfirer. Although Fir Trees was only a small pit, it constantly broke the output records in the area. Gordon remembers the time when Gordon Davies, who was working at Fir Trees, broke into the old Habergham Colliery workings, or as it was known locally, 'Cheapside Pit'. Gordon Hudson at this time was working on supplies at the pit. The old roadway was all brick-arched and still in good condition – there were even the old coal tubs still in place on the rails. These, though, had all rusted away leaving just a skeletal framework on wheels. The old Habergham workings were inspected, but they dared not go too far into them in case of gas – blackdamp.

Gordon recalled a few of the firemen at Fir Trees, including 'Jacky Red' because of his red hair, and Freddy Moreby. Another was Vinnie Anderton, who had worked with Gordon's granddad at Towneley Pit, Towneley Park Pit and Salterford Pit. Vinnie was the only man at Fir Trees who could use a longwall cutter, so when one was installed at Fir Trees he was naturally put in charge of it.

Most of the men at Fir Trees at this time used to get changed at Bank Hall Pit, and then they were taken back to Fir Trees on an NCB bus. The poor unfortunate colliers at Padiham, though, were sometimes taken to the pit on the back of an open NCB lorry – dust and all. Gordon Hudson still had his trusty motorbike however. The men who travelled to Fir Trees from Bank Hall were also taken back there to get changed and showered – Fir Trees had no showers of its own for years, although they did eventually get some.

Gordon Hudson was injured at Fir Trees Pit while unloading some 'Dowty' props off the chain conveyor when the prop caught on one of the flights on the conveyor. This sprung up and caught him on the skull and fractured it. He was placed on a stretcher and taken out of the pit – but he remembers nothing of this, only waking up in hospital with a sore head. He was in fact very lucky; had the prop not knocked out him of the way, he would have been pinned to the roof and almost certainly killed. Gordon recalled a Padiham collier who was using a pneumatic air gun to bring some rock down when he hit an unexploded fuse used for shotfiring. The detonator exploded and almost killed the man, all his body was covered in coal dust, wounds and blood. He did survive though and came back to the pit later on. 'We had a fireman named "Taffy",' recalled Gordon. 'He was not liked too much because he would put you on work which was either harder or should have had more pay, but he would not book you in for it and you lost money.' Gordon enjoyed his time at Fir Trees Pit: 'It was a good little pit, where everyone knew each other – but I did not like it at Bank Hall.' He went on to say that, 'I enjoyed pitwork that much, I would have done it for nowt'. Gordon came out of the pits during the period when the local collieries were being closed down and went into bricklaying. Today Gordon enjoys his days doing what he likes best: 'just taking it easy'.

Today there is no trace of the old Fir Trees drift mine, it has all been landscaped, however the more observant might just notice a board on a nearby telegraph pole which states 'Fir Trees Colliery'. A silent memorial to a past industry – everything else has gone.

7

Hapton Valley Colliery

The history of Hapton Valley Colliery, near Burnley, is really in two parts – that of the old Spa Pit and the New Pit. The old Spa Pit was located near the bottom of the present-day Municipal Cemetery at Burnley and dated from around the mid-1850s. Work on sinking the New Pit, which consisted of the No.3 shaft which was 18ft in diameter and 514ft deep, and the No.4 shaft which was 12ft in diameter and 488ft, was completed in 1910. However, development work and other problems meant that these shafts were not fully operational until 1932. After the nationalisation of the coalmining industry, the New Pit a underwent massive reorganisation. New surface and shaft bottom tub circuits were installed and larger tubs introduced. The East Side Drifts were completed, opening up a whole new area of coal beyond the Deerplay Fault.

Three years later, in 1961, the arguments for a surface drift at the pit came to fruition when work was begun driving an inclined tunnel from the surface. The new surface drift 'holed' into the existing workings at the bottom in January 1962. In order to find out the effects on pressure and ventilation, experiments were carried out running both the main fan and the booster fan together and individually. It was considered that the booster fan at the bottom of the return airway in the East Side Drift was no longer required, and it was taken out of use and put on a care and maintenance basis. Two months later the pit was thrown into the national headlines by Burnley's most serious coalmining disaster. On the morning of 22 March 1962 a methane gas explosion seared through the workings of the Rise Two District at Hapton Valley Colliery. Sixteen men and boys were killed instantly, another three were to perish over the next few days, and many others were seriously injured. Almost fifty years on it is heart-warming to note that those who died that day have not been forgotten. Each anniversary of the disaster is remembered by large crowds of former mining comrades and family members who walk to Burnley Cemetery for a remembrance service there.

Another view of the layout at Hapton Valley Colliery, probably taken from the roof of the pithead baths. One of the winding wheels on the upcast shaft seen in the foreground was later used on the mining memorial which is now in Bank Hall Park. (W. Rawstron Collection)

The pit never really recovered from that dreadful day, but it did become a profitable pit for the National Coal Board, mining coal from the Upper Mountain Mine and the Union Mine. However, time was running out for Hapton Valley and the closure of the pit was announced in February 1981. The last shift at the pit was worked in July the following year, just a few months after the twentieth anniversary of the disaster there. The comradeship, skills and bravery of the mining communities also went with the closure.

Robert 'Bob' O'Hara	
Age at interview	*89½*
Years in mining	*c.35*
Collieries worked at	*Kingshill Colliery, Allerton, Lanarkshire, Reedley Colliery, Burnley, Wood End Colliery, Burnley, Huncoat Colliery, Huncoat, Calder Colliery, Simonstone, Bedford Colliery, Leigh, Astley Green Colliery, Manchester, and Hapton Valley Colliery, Burnley*

Hapton Valley Colliery

Hapton Valley Colliery, near Burnley, just after the completion of the surface drift on 19 January 1962. Just three months later the little colliery was thrown into the national headlines when a methane gas explosion seared through the Rise Two District underground. Nineteen men and boys died and many others were injured. (Jack Nadin)

Here Bob recalls his early childhood days in Morningside, a pit village in Lanarkshire, Scotland and his lifetime in the pits. Bob is obviously one for great detail: when asked his age he gave, not 'eighty-nine', or 'nearly ninety', but eighty-nine-and-a-half, rather like 'Adrian Mole aged thirteen-and-three-quarters'. He describes the poor housing conditions at Morningside village, a situation noted by the authorities in 1909, eight years before Bob was even born. When Bob was born, the atrocities of the First World War were still wiping out thousands of young lives in Europe; in the home country social conditions were also extreme, as Bob is about to tell us.

I WAS BORN the middle child of twelve children, on 9 January 1917 in Morningside village in Lanarkshire, Scotland. My parents were John and Margaret O'Hara, sisters were Bella, Sarah and Mary, and my brothers were John, William, Archie, James, Tom, Andrew, Dan and David. Morningside was an all-mining community, with the exception of a few railway workers. It was a small village with a population of around 1,000 souls – although we

Bob O'Hara studies the masses of official documentation relating to the inquiry into the Hapton Valley Colliery Disaster which he has in his possession at home, and remembers with affection those who died and were injured that day. 'I still have visions of what happened that terrible morning: it was the worst day of my life,' he said. (Jack Nadin)

did have two railway stations. One went to Edinburgh, the other to Glasgow, and they were used to transport the coal and iron – a passenger service was also available to those who had enough money for the fares. The house I was born in was one of the rows of housing built to accommodate miners from the local collieries. Our particular row had twenty-four houses, held up at the back by brick-built butts to prevent them falling down. The mine workings at the rear had caused considerable subsidence. The house had two rooms on the ground floor, no upstairs, and the parents slept in the back room and the rest of us in the front. The front room had two built-in openings with built-in beds. We slept four in a bed 'head to tail', two at one end, two at the other end. There was no running water in the house, at the end of each row of houses there was a standpipe, below which was an open gutter made of half clay pipes, and there were no toilets. Also at the end of each row of houses there was two brick-built middens used for rubbish and ash as well as the contents of the buckets which were used as toilets during the night. These were emptied maybe once a week and spread on the local farmers' fields.

If any money was available, we might have oilcloth on the floor and a handmade rug made up from old rags cut up and pegged through a piece of sack cloth. It is quite understandable that in these cramped and confined conditions disease and illnesses was rife; measles, chickenpox and other contagious outbreaks were common. Heating was by an open coal fire with a built-in oven for cooking and baking in. This also had a swinging arm to hold and carry large cooking utensils.

The main diet consisted of porridge every morning and sometimes after school. Other foods were broth soup and potatoes boiled in their skins, and

Typical miners' accommodation in Lanarkshire, Scotland, similar to those described by Bob O'Hara. (Jack Nadin)

bread and jam. As it got nearer the weekend the kids would have to forfeit their bread and jam in favour of the menfolk who worked down the pit. The porridge was made at night using fine oatmeal and kept warm in the oven until breakfast time so that the workers could set off at 6 a.m. to the pit – they might also have a piece of toasted bread. It was an extremely hard life for the mothers of the families; they would have to be up each morning at 4 a.m. to get everything ready for the pit workers. With this coarse lifestyle and hardship life expectancy was short, sometimes very short. My mother died aged just forty-two years old – Father died aged sixty due to silicosis, the dreaded miners' disease. Most miners in Scotland wore hardwearing thick moleskin trousers: these were washed and dried round the fire overnight to be ready for work in the morning. How the women coped with washing these clothes has to be marvelled at. It was not unknown for these trousers to stand up stiff when the men came home on a frosty morning after night shift. There were no pithead baths around here until 1929, so everyone had to walk to and from the pit in pit clothes.

When I started at the school, Morningside School, we had a slate to write on for sums and learning to write. Our shoes, actually boots with steel cladding, had to be spotlessly clean or we got the leather belt. Money being in short supply, clothing was limited and usually handed down from the older boys. I cannot remember wearing a coat to school, but I suppose I must have. During the eight weeks' school holidays we mostly ran about in our bare feet, consequently the tops of our big toes flopped about. We used to play football, and inventive as boys are, we used the middens at the end of the row as goals. The most common ball was made up of several layers of paper, and then wrapped into the shape of a ball, unless some well-off boy had a tennis ball. When the ball landed in the midden, there was usually a row about just who was going to get it back out again. If it landed in the wet stuff, such as the contents of the toilet buckets – then the game was usually suspended until another ball was acquired. Sometimes during the school holidays we would spend a day damming up the small river for swimming in and have a picnic around the sides. The picnic consisted of potatoes roasted over a fire and a tin can to boil water for a brew of tea. To eat the spuds we had to knock off the black burnt bits to get at the inside.

At the time of the 1926 Coal Strike, I was nine years old, and now wonder how the mining communities survived on the little food they had. The Co-op gave support for the miners' families by donating food, and sometimes a limited amount of credit to be paid back when the strike was over. There were also soup kitchens manned by the people of the village. I can remember going to the soup kitchen with a white enamelled bucket and getting it filled, along with a loaf of bread for the family of eight children and parents. Father was a committed trade union official and had to pay fines over his conflicts with the strike-breakers and his confrontations with the students used by the police to break up demonstrations. These students were mainly from the upper classes who would normally attend university. I also remember the women of the village going to the pit and walking home with the strike-breakers shouting 'Scabs' and 'Dirty Buggers' all the way home over a mile away. The chanting also continued for some time even after the blacklegs got home, but it did not deter them. There was always the police with them travelling to and from work and near their homes in case any hothead tried to stop them. It was learned later that they were all members of the lodge of a certain society. There were two of these blacklegs in our row of twenty houses. After the strike the men were rewarded by being made contractors, able to give work to the men who were then paid by the contractor. This was called the butty contract system.

I was aged fourteen when I left school, there was no school leaving certificate then. My first job was in a newsagent's shop delivering papers in the morning and at night. Because this was a rural situation, the morning delivery lasted about two hours, and I must have walked four miles a day. I got no pay for this, because this shop was my father's shop. I did this work for two years while waiting for a job at the pit. The system was that when you left school you had to wait for a vacancy at the pit. When I was nearly seventeen I got set on in the picking rooms separating the stones from the coal. The pit was at Allerton, a few miles from Morningside, and called Kingshill Colliery. Kingshill No.1 at Allanton closed in November 1968 and Kingshill No.3 at Allanton closed in July 1974, bringing an end to coalmining in that area. [At the time Bob was working there these pits were owned by the Coltness Iron Co. Ltd, Newmains – Ed.]

The working week was forty-eight hours, eight-and-a-half daily, Monday to Friday, the rest done on a shorter day, Saturday. Those were the official hours although we were actually at the pit nine hours a day with a half-hour break. The wage for this work was 1s 11¼d. This would be equal to nine and a half new pence in today's coinage. We had 11s 11d weekly, which would be equal to just under sixty pence a week today. I got half a crown weekly as pocket money. My next job was underground working a hand pump – I had to use a flat bogie to pull myself through water by using the roof girders and manually operate the pump. I spent the whole of the shift on my own in the main return airway, almost half a mile from the working production face. For light, I had a carbide lamp, and this job took a lot of getting used to. The fireman would visit me twice during the shift to make sure everything was OK. My next job was as a supports supplier, taking props and other essentials to the colliers on the coalface – at last I was back with companions. This work involved pushing flat bogies full of props to the return end of the face. The props were sent through the face on the conveyor belt as the colliers required them. My weekly wage for this work had risen to 18*s* and my pocket money also went up to 5*s* a week. It was during this time that I witnessed the first fatal accident at the pit. When I was sending props down the face, a shout came back to tell the fireman that Daley Bill had been killed through a heavy fall of sandrock. This was a real shock because I came into the pit in the same mine car as him and two other colliers. We had to travel the mile-and-a-half on a train of empty mine cars – there was no manriding facilities at this time. Daley Bill was a happy and friendly sort of person and would sing all the way into the pit. This was 1934, and the songs he sang at that time were 'Sweet Meat Joe', 'The Candyman' and 'The Daring Young Man on the Flying Trapeze'. Thinking about Daley Bill singing happily away on his way into the pit, and then getting killed so suddenly, had a profound effect on me – something I will never forget!

This coal here was called the 'Wall of Death' because of the terrific breaking noise it made when having what was called a 'weighting'. If this weight hung for a long time before breaking off, perhaps days, there would be a huge roof fall all at once, and the weight could be thrown forward on the supports on the coalface. The colliers had a second sense about this 'forward weighting' and would listen to the noise made by the wooden props splitting under the load and stress – they knew when it was time to get out into the nearest roadway before it all closed up. When the men were supplied with steel props to cut down on timber, they took some convincing that this was a good idea. The men were used to the timber props warning them of impending danger, steel props though did eventually come in. My next job was turning the mine cars around at the delivery end of the conveyor belt collecting the coal from both sides of the coalface. The air here was thick with coal dust with the coal being delivered into the mine cars – there was little or no dust suppression by using water in those days. My wage for this work was 25*s* per week. Each longwall face at this pit was 120 yards long, and the coal seam averaged between 30in to 36in high. Following this I was placed on roadway repairs, this is when my wages jumped to £2 5*s* per week and my weekly pocket money rose to £7 6*s* a week. After this I was put on ripping, driving the roadways at 20-yard intervals along the line of the coalface. They called these 'dummy roads'. The idea was to form packs made of stone to support the roof and to encourage the roof to collapse between these packs and take the weight off the coalface. It was in 1938 that my father died from silicosis. No compensation in those days. I was twenty-one years of age at that time and had to take over as 'man of the house' to keep the family together. There were six younger brothers, and my sister who was nineteen, running the home.

We got 7s 6d for each brother under fourteen, and 9s for those of working age in dole money. It was about this time that we got our first paid holiday week, although we had to pay for it ourselves as they took 1s 6d out of your wage each week. I carried on as a ripper until the spring of 1941, when I decided to go down to England to join the army at Halifax – we were refused because we were miners. I tried the RAF but was refused again, they told us we needed permission off the colliery where we worked, and so we set off back to the pit. Fortunately the manager said he could let me go, being a mining community there was plenty of labour in the village. I joined the RAF as a ground gunner. I tried to get in as air crew but wasn't given the chance because I had not been to grammar school – I was in the RAF for four-and-a-half years, three-and-a-half years abroad and a year in England. I landed back in England in June 1944 and was billeted in Blackpool, where I met my wife Irene, and in November 1944 we got married at Burnley. After some weeks in Blackpool I was posted to the south of England to be ready to be posted to the Continent – I did not go until early April 1945 and spent the next six months in Germany until I was sent back to the coal mines. I consider myself lucky that I was not involved in action during my four-and-a-half years in the forces. When I was demobbed in October 1945 my wife and I moved up to Scotland because I was still officially the council tenant of the house at Allerton the family had moved into in 1938. I had taken over the responsibility of the five children who were aged from ten years old to sixteen years old. When I look back the responsibility must have got to me, this may have been why I joined the forces.

Irene, my wife, was having a difficult time with a sister and one of my brothers, so we decided to go back to her home town of Burnley, Lancashire. We had a country cottage on the edge of Walverden near the town of Nelson – but it was like going back in time, back to pre-1927, there was no water, electricity or gas. Water was obtained from a trough supplied by spring water, during the winter or during a hard frost we had to go into a field for water from another spring. I got a job at one of Hargreaves collieries, Reedley Colliery, but to get to work I had to walk a mile just to catch the bus to the pit.

I was working as a repairman making good small roof falls in the roadways; the pay for this was £4 16s per week. In 1949, we managed to get the tenancy of a council house in Haggate, and this at least cut out the mile walk each way to get to work. The men at Reedley Colliery and the

Reedley Colliery downcast shaft. If ever there was a symbol of an industry, the colliery winding headgear would be the image that most would recognise instantly. (W. Rawstron Collection)

Wood End Pit were the friendliest people I had ever come across anywhere; I was welcomed in every way. While at Reedley Colliery, I started studying at Burnley College in colliery engineering and management and after four years I gained a second-degree certificate and under-manager status. While doing these studies, I did several jobs in the pit – prop withdrawing, ripping, and then as a fireman. The dangers of mining were revealed once again when the Reedley Colliery witnessed three separate fatalities while I was there. I later moved to Huncoat Colliery as an overman for a year, and in 1957 became the under manager at the Calder Colliery near Simonstone on the outskirts of Padiham.

Both the Huncoat and Calder Pits were under the charge of Bill Oldroyd. While I was at Calder a great deal of money was spent on skip winding and

Happier times. Bob O'Hara with his wife just after they were married in Blackpool. (Bob O'Hara)

opening up new coalfaces, but this was also a time of pit closures and a fall in the price of coal and in July 1958 Calder Colliery was closed down. The men employed at Calder Pit were transferred to Hapton Valley, Huncoat and Thorneybank Collieries, while I was moved to the Fence No.1 Drift Mine near Fence village and Fir Trees Colliery, another drift mine close by. The strata at these pits was very brittle, and because of this the props had to be set very close together. Besides the obvious difficulties of this, another problem was also raised when a large collier was taken ill and had to be carried away from his workplace. This involved taking a prop out to move him forward a couple of feet, and then resetting it, and then repeating the process until it was possible to get him out of the pit by way of the main drift. The seams were so steep at Fir Trees that any tool you were using had to be placed behind a prop otherwise it would disappear, sliding down to the bottom of the roadway.

The coal at Fir Trees was got by hand pick, the colliers were paid by how much they had advanced forward in the seam. The advance was measured in feet and to the nearest inch and each collier's place was measured daily. The seam, the Arley Mine, was 4ft 3in high, much of the roadways were 4ft 6in high. Because the seam was so steep, machines could not be used, not even pneumatic gun picks – the coal though was the finest coal ever mined in the Burnley Coalfield.

In 1960, I was moved from Fence No.1 and the Fir Trees Pit to Hapton Valley Colliery with a position as under manager. Hapton Valley worked the Mountain Mine and the roof conditions were entirely different to the Arley Mine at Fence and Fir Trees. I was amazed to see the colliers cleaning up all the loose coal from the undercut before even bothering to put up supports. They were breaking every rule in the support book! They were filling out their allocated yardage as soon as possible, once this was done they were allowed out of the pit. I came across several of the men on their way out of the pit at noon, most of them heading straight to their local pub. They would obviously be ready for a pint or two after shifting 20 tons or more, and then setting the props. I was at Hapton Valley when I had the worst experience of my life – this was the day of the disaster [22 March 1962] in which sixteen men and boys perished in an explosion and a further three died from injuries – a further thirty-one men were injured, many of them seriously. When the

The force and the devastation of the terrible explosion at Hapton Valley Colliery on 22 March 1962 can be clearly seen in this photograph. Broken, overturned tubs and bent metal record the aftermath of the disaster that claimed the lives of nineteen local miners and injured many more. It is gratifying to note that, even after all these years, the memorial services to the victims are still well attended – those who perished that day have not been forgotten. (Bob O'Hara)

explosion occurred, I went down the pit to stop the return airway manrider from being used to bring the men out of the pit in case of dangerous gases in the return airway. This was about 9.50 a.m.

I had received word from Jamie's Junction about the air doors opening and closing twice. This was just before 9.45 a.m. when I received a message that there had been an explosion on No.2 Face. I was still on the pit top at this time because the stand-in manager, Bob Kennedy, asked me to stay back when I was going down at 8.45 a.m. This was because he wanted to discuss the weekend work – this had to be sent to headquarters on the Thursday. When he got word of the explosion, he asked me to station myself at the bottom of the No.2 gate to organise the transport of blankets and stretchers for the injured men. The Safety in Mines people had gone up the return airway to declare the safe movement of the injured. From my position I had to report the names of the injured and non-injured to ensure that all were accounted for. I was very much impressed by the way so many men made themselves available, even men who were on different shifts were coming to the pit to help their comrades. These were very good men to use and have in an incident like this. It was extremely hard going up to the coalface and carrying stretchers back to get the injured transported out of the pit. Some made several journeys in a very difficult roadway and not the best of footings.

The area nurse was Sister Maude Waggot. It must have been some experience for her to go up to the coalface and administer as much first aid as possible. I had told her to go out of the pit, and then she told me who she was. She certainly did a good job. I can declare that all those volunteers deserved a medal. At 5.00 p.m. I had to go to the bottom of the return airway to try and identify those who had perished – an extremely difficult job. In some cases I could only do this by using lamp numbers and check numbers. The victims were then moved to the pit top where the baths attendant, Jimmy Farrer, cleaned them up. I came out of the pit with the men who had carried them from the coalface to the bottom of the return airway It is gratifying to note here that these men had been working from 10.00 a.m. until 6.00 p.m., they had toiled relentlessly in what must have been the longest day of their lives, it must have been extremely trying for them. It was not possible to start the pit working again for several days, and most of those days I spent going through the explosion area with the Inquiry people from the Safety in Mines department. I was with them in order to tell them about anything they found; it was a very through investigation. I was back at work for production again after this period. The next few months until the Public Inquiry in June consisted of extreme mental anguish – my doctor advised me to take sick leave, but I refused it. I carried on at Hapton Valley until early in 1963, and then gave notice of leaving the mining industry. From 1963 through to 1964 I worked as a lagger and general worker. I then left to work as a tunnel foreman for a private firm at Bedford Colliery, Leigh and the Astley Green Colliery, Astley Green, Manchester. I left this work in 1965 and began work at Huncoat Power Station, I was then asked to go back to Hapton Valley as materials control officer. In July 1979 I retired from the work on the early retirement scheme. I have been retired now for twenty-nine years – and I consider myself fortunate in being free from serious injury or mining breathing diseases from my thirty-five years down the pit. One final note: I have lived in the Burnley area now for fifty-nine years, and I have nothing but admiration and pride for the people of this area.

∞

Robert 'Bob' Woods
Age at interview 82
Years in mining 42
Collieries worked at Hapton Valley Colliery, Burnley

I met Bob at his cosy little home on the boundaries of Burnley and Padiham for the following interview in October 2006. I had to admire Bob's ability to recall his coalmining days in great detail. What follows is a transcript of that interview.

I WAS BORN [on] 14 June 1924 on Cog Lane, Burnley at No.212; the house has now gone. They were terraced houses, my dad could have bought them all, eight or nine of them for £80 apiece – but we did not have the money. The houses had big cast-iron fire ranges in and we used to keep our pit clothes in them, in the oven side, to dry them out ready for the next shift the day after. It was all fields around there at this time, the Stoops Estate had not been built yet, and it must have been a mile, or a mile-and-a-half, to the pit. It was bad in winter: we used to put newspaper round our legs over our pants and tie them with burning band to try and keep them dry. When we got to the pit we used to go into the boiler house to try and warm up – because there were no showers then. We got bathed at home, oldest first – by the time it got to me the water was thick. We had some bad winters then, and the snow used to pile up as high as the cemetery wall down the ginney track. Many a time we used to have to come up the pit and clear all the line of snow to get the tubs going again. My father was Robert Woods, same as me, he was also a coal miner at Hapton Valley, and my mother was Elizabeth Woods. We did nearly a hundred years down the pit between us, me and my father. He

Bob Woods in his back garden in September 2006. (Jack Nadin)

also played for Burnley Football Club. I went to Hargher Clough School and then Rosegrove School, and I left there when I was fourteen in 1938.

I started at the pit when I was fourteen years of age; my father made me go to night school. I was working in the pit until half past two or three o'clock and I had to be at night school at six o'clock, and I dare not miss, because Father was the gaffer o'er me then. My first job at the pit was tub oiling, but I only did a week or two. Then they wanted a 'tally lad' and picked me, because I was the only lad with a bike. You had to have a bike. I will tell you what I used to do: first thing at morning, this tally tub would come off ginney with all the previous day's checks in a big tin box. I used to pick them up and go into the tally hole then and hang them all up; every collier had his own checks. They got paid by how many tubs they filled, because they were all checked. I did this job for about twelve months, maybe two years. Part of this job was to go for fags for the lads, all the pit toppers. There was a shop in Valley Gardens and she used to have all the cigarettes and toffee and all that. I used to go there every morning for cigarettes, Craven A, Woodbines, and all that, it was all part of the job. The shop was the first house you came to in Valley Gardens, they sold toffee, chocolate and all that. By the time I had done all that, the manager had come and I had to go into the office and get all the letters off him and find out where they were going and take them to the drift behind the Hapton Inn. I had to cycle up to this drift with the letters, hence the need for the bike. I got a bit cheeky then, I got to know a lorry chap who went about eight o'clock or half past eight. Every morning he was on that run so I used to hang onto the back of his lorry on my bike all the way up the hill. Later I went down the pit. There was no face training then, you did so many years drawing with the tubs and then you went on what they called 'your two years'. Before the war, you were all on your two years, the best, they were, in the Territorial Army, but as soon as war was declared they were conscripted first, but no bugger come back hardly though, a lot of them were killed. So we had to go straight onto colliers' work then, without going on the two year.

I tried to join up, but you could not get out of pit even if you wanted – they would not let you because of the war. I went to Liverpool Docks to try and get on, but the soldiers there would not let me aboard. I spent two days there, no food and no lodgings. I then joined the Home Guard. I was working in the pit all day and going to the Home Guard at night and college in-between: two or three nights with the Home Guard and two or three nights at college. The manager at Hapton Valley when I first started was Harold Warne, in later years it was Adam Weir, 'Jock' Weir we called him. When I first went down the pit I went on drawing little tubs coming off the coalface in the Lower Mountain Mine. As you stepped out of the shaft at the bottom you went round to the West Side – Low Bed and all that. It went on for miles. We went to No.11, and then we had to go through the step, a fault, and that were hard work, trying to find out where coal was. We had to make a face off then, and then we carried on again. I was with guns, what they call pneumatic picks. We did not tram in there, but we used to tram into the West Side. We used to go there a time or two when it were flooded out. The coal at this time was machine-undercut and shotfired at night, but then they got these pneumatic picks and you just had so many yards apiece, about twenty of us. The coal was not undercut then, we just brought it down with the picks. My first week's wage at the pit was 18s 7½d for six days. Then at half past two I had to finish all my work, come out of the pit and then go into the lamp room cleaning lamps, filling them up, topping up the oil lamps. I was doing that until four o'clock, I got paid extra for that, 1s 1d. I had to rush home then, get washed and have my tea. My mother used to work in the UPC tripe shop on Fleet Street in the town centre and I used to get my tea there, steak pudding and chips usually. I had to go then to the college on Ormerod Road to learn practical mining. You had to pass every year or else you had to pay, so I had to pass, or my dad would have gone mad. But when I got to about nineteen or twenty years of age I was more for the girls and snooker. We used to go down town on the 'Drag' where all the girls walked up one side of St James's Street, and the boys walked up the other side and swapped over at the top. I also went into Burton's snooker hall – I was mad on snooker – and I went down Carlton snooker hall that was downstairs. Sometimes we would go to the Empress; there was skating there at afternoon, and dancing at night.

I asked Bob if he had had any accidents at the pit.

I broke my back, I did not actually do it in the pit – it were through pit though – all my bottom vertebra had levelled out. I was at home, just throwing something in fire, when I got this pain – I was paralysed. This was through shovelling down pit, I didn't get halfpenny, because it happened at home you see.

I once broke my foot. We would pull the tub out when it was full off the haulage. I had pulled two or three out when one run over my foot. Nobody bothered, they just pulled me to one side, because as soon as they were done they could all go out of the pit then. I had to go out on my own with a broken foot. I was once working with David Allen, I was drawing, and someone came running off and said 'Someone's been hurt or something at ginney'. You see sometimes the tubs would jump off road and the ginney tenter used to go under and put it back on road – that's when it must have happened, Bentley I think they called him, Dicky Bentley's cousin. The tub must have caught him, he was half and half over tub – he was killed.

When I was thirty-three years old, my wife's father died from tuberculosis, so we all had to go and get an X-ray and I was called back to see the specialist who told me I had dust on my lungs – dust on my lungs at thirty-three years old! Going to Manchester to see another specialist confirmed that I had silicosis, and they brought me out of the pit right away. I only did about six months out of the pit, all one summer. I had no wage and two kids to fetch up. So I decided to go in for a deputy certificate to get out of the dust a bit and when I came to go for this I practically knew it all through going to college beforehand. So I passed that with flying colours. When I put in for deputy I had to leave for thirteen weeks and go to Huncoat Colliery for two or three days a week and go to the college [the] rest of the week.

When the explosion happened on the No.2 face I was on that face, I was shotfiring on that face. One of the shotfirers was on his holidays, so they asked Jack Halstead if he would go shotfiring on nights – he was on another face, No.5, and he said no, he was not coming in to go on nights. So

Hapton Valley Colliery bowling team in the late 1950s. From left to right, back row: Harold Yates, Robert Woods (junior), Frank Wayward, Jack Woodheap, Joe Ellywell, Johnny Hargreaves. Front row: -?-, John Henry, Robert Woods (senior), Harold Warne (colliery manager), Tommy Dulsey. (Bob Woods junior)

I think it was Bob O'Hara who asked me to go on nights for a week, so I went on nights, and that's when it happened, the explosion, on the Thursday morning.

I was not there when it happened, because I was on nights and had just come home. I had just had my breakfast and got in bed when my wife came running in and told me about it. So I rushed out right away, and I got to the lights at Rosegrove and there was a car passing so I pulled him down. He stopped and it was a reporter. He said, 'I am going to the pit'. He'd heard about it and all, and he were going to pit. So as soon as I got there, two or three of them were there, and there was Sam Bullen's wife, and she said, 'Can you find out what's to do with our Sam?' So I went straight in – I got changed and went down pit. And then Jimmy McKillop came – he was on afternoon shift I think. Eric Watmore came then, the manager at Bank Hall Colliery; the under manager at Bank Hall, Mr Rawstron, was working as stand-in manager at Hapton Valley because Adam Weir, the manager at Hapton Valley, was down London somewhere on a course. After the explosion Eric Watmore took over things. There was myself and Jimmy McKillop in charge of bringing those who did not survive out of the pit. We had to try and identify them, who they were, and Bob O'Hara was on a phone about a mile further down. The main job then was to get the survivors and wounded out of the pit. Jimmy McKillop, Eric Watmore and myself were a man short, and then Billy Ellins came up, and we got Chris Brown on a stretcher – he was the only one who wore clogs, that is how we knew him. He was the last of those who perished in the explosion to be brought out. We did not bring them out through the drift, because it was crowded with relatives and newspaper reporters, we brought them out by the shaft. When we got out of the pit, I said to Jimmy McKillop, 'Come on, let's go and have a cup of tea' – we had had nothing to eat all day. But when we got into the canteen it was packed out, you couldn't stir, they were all there supping and eating all cakes as well. So I said, 'Come on let's go and get changed'. I am not sure whether it was Bob O'Hara or not, but when I was coming across pit top after getting changed someone shouted from the top of the lamp room ramp, 'Don't forget, come in to work in the morning Bob'. I could have strangled him! But when I got home I realised that I had to go back in, so I turned up the day after – but all the men had the week off. I was there, I had to be there, and I had to go back down the pit every day. The following Wednesday morning I had to go down the pit and fire the shots that were left in the tailgate heading, there was about five or six left, and I had to fire them. It was definitely not shotfiring that did the explosion because we found Jack Halstead way back away from the heading, and his shotfiring battery was still on top of the boards near the caunch and the coalface. So he was coming towards it at the time, so no, he had not fired. So what we put it down to was the gob, this had gone about 30 or 40 yards and it all must have all come in and fell at once and released a load of gas. Then when it got into the tailgate it layered on the top because it is lighter than air and it must have got down to where those lads were with the tackle tub. Then it only needed a spark. No, it definitely wasn't shotfiring.

Even after all this time, it was clear that the events of that day were still clear in Bob's mind and that it was still quite emotional for him, so I changed the subject here, and asked about the two times Hapton Valley Colliery was flooded.

Yes, I was there both times, one time the pit had flooded and we had to start getting coal anywhere we could, all around about the shaft. There was a creeper that went up and brought the empty tubs down and then they went onto the ginnies to be filled at the face. They started a face right above there and it lasted us a few weeks and then we must have broken into some old workings dating from the 1800s and they were full of water. They were un-charted workings, no one knew about them, the water gushed out. Then the other time, Ben Rushton and I found it; we were working on Sunday night, me and Ben Rushton, and we had to go around all the faces. We got to this No.2 face and the water was gushing out of the floor. It was a mess, and that flooded the pit out altogether. The water filled up to the bottom of the drift and it stopped the air circuit, so they had to keep pumping a little bit to let the air flow out, and the gas kept building up. In the end we had to get those special submersible pumps in. These were working all the time after that until the water went away. This was the first time that the pit flooded, all the pit was stopped and we had to get coal from anywhere. We started another face up about 20 yards down the drift. The main job was to get a bit of coal anywhere just to keep up production. The pit was stopped for weeks until they got that water out and the pit went back onto full production.

I asked Bob if he ever took any personalities down the pit, such as town representatives or 'Coal Queens'.

I don't remember any of those. But when I was working weekends – we had to take our turns at working weekends then (we did not get anything for it) – I used to take one or two down, one or two of my mates. I took Lol Yates, 'Butcher' they called him down the pit, and his brother. I also took Barry Towneley down. They wanted to go down the pit, so I took them – it was all unofficial of course. When I took Barry Towneley down, I started the coal cutter up; you should have heard him shouting about all the dust and noise, it frightened him to death. That was the last time he went down the pit. I enjoyed my time working in the pit. The last few years though I spent working on the surface because of my back injury and my chest complaint. I finally retired from the pit about twelve months before it finished. They were going to finish it, but there was one face doing good, so they decided to keep it on until that finished.

    I enjoy my retirement and take things easy now, we like to go away as often as we can on holidays. In the first few years I tried getting a part-time job, but then they started taking tax out of it, so I said no way and packed it in. I just relax now and enjoy myself, I think I deserve it.

Archer Lee
Age at interview 79
Years in mining 25
Collieries worked at Bank Hall, Burnley; Hapton Valley, Burnley

I have known Archer Lee for most of my life, including the time I worked at Hapton Valley Colliery. Friends made at the pit lasted a lifetime – and even though the colliery has been shut down for many years now, Archer and my uncle Terry Pickard are still great friends and regularly play golf together. Here, Archer was able to recall his early life and his time down the pit with great detail.

ARCHER LEE WAS born at Malt Kiln Street on 7 April 1927 in the old Hill Top district of Burnley, the third son of Archer and Winifred Lee, *née* Grogan. There were five other sons born to this couple, although unfortunately the eldest, Raymond, died as a result of an accident while playing on his bike. The accident left him with an injured hip which eventually caused his death at the tender age of fourteen – he had only just left school. The other children were John, then Archer, Peter, Henry, then Albert, the youngest. Besides bringing up their six sons, Archer's father and his mother Winifred also brought up Winifred's brother Martin, who was orphaned when their

Colliery overmen Jimmy Holden, Mick Hargreaves, Archer Lee and Ben Rushton enjoy a pint on a trip to Morecambe about 1960. (Archer Lee)

father died early through working at the coke and gasworks off Parker Lane. The fumes off the coke ovens caused many of the workers there to die young – not many of the workers there lived into their forties. Martin began work at the old Towneley Colliery at the age of fourteen and was there until it closed in 1948. He then went to Hapton Valley Colliery until he retired. Martin was at Hapton Valley at the time of the 1962 disaster. He was a great mate of Chris Brown, both had worked at Towneley before coming to Hapton Valley, and Chris was one of the nineteen who were tragically killed.

Malt Kiln Street itself, off Church Street, near to where the area's first Sainsbury's store was built, took its name from the old kiln which still survives. It was an area of blackened stone mills and factories, narrow backstreets and crowded housing. As many as eight to ten houses all had to join at just a few outside 'long drop' privies on the back street. Archer's father worked at Grimshaw's Brewery just across from Church Street, and although the late 1920s and early 1930s were grim days, they never missed a meal. The Hill Top district however was earmarked for demolition by the council, and the family moved to 17 Tiber Avenue on the then-new Stoops Estate in 1933.

Archer Lee and his brothers were all good sportsmen. Here we see the Hapton Valley Colliery Works League Champions team of 1952-53. From left to right, back row: T. Greenwood, Frank Heywood, Alan Harvey, Dick Bentley. Middle row: John Wilde, Brian Hoy, John Barker, Joe Walsh, Derek Redford, Henry Lee. Front row: Harold Warne (colliery manager), Albert Lee, Jim Hurley, Archer Lee, Lol McDevett, Neil Hurley, Norman Stowell. (Archer Lee)

The children, including Archer, were educated at St Mary Magdalene's School which was on Halslam Street near Gannow Top. Archer left school at the age of fourteen in 1941 and started work right away at Bank Hall Colliery off Colne Road. There was no training in those days. For the first few weeks you were placed in the screens taking the muck and stones out of the coal as it moved past on a conveyor belt. Archer then became a 'tally lad' at the No.4 shaft at the pit. Colliers then were paid by the tubs they filled, and to mark their tubs they hung a tally on the side. In this way their tubs could be identified and the collier would be paid for that tub. Archer's job, as a 'tally lad', was to collect the tallies that had been taken off the tubs and sort them out ready for redistribution to the colliers. Once the tubs had been weighed by the checkweighmen, Tom Whiteley and Bob Emmett, they were emptied into the screens and the tallies thrown in a bucket. Archer would collect this bucket and replace it with another one and take the tallies back to be hung on a board. There were three faces working at Bank Hall at this time, Cox's Rise, Dip One and Dip Two. After six months as a tally lad, Archer was put on prop running, taking props to the coalface along the gates, or tunnels, with Terry Barber. These tunnels were just the height of the coal seam, perhaps a bit higher, 4ft at most. Sometimes Archer would go drawing for the rippers, Jimmy Naughton, Charlie Barrett and Ted Whittaker, who were opening a new face to be named Joe Wright's Face, and he would fill the tubs brought down by the rippers with stone. The stone itself was brought down using pneumatic picks. Once the tubs had been filled they were taken out of the pit. Because this was an inclined roadway, Archer would have to put two 'sprags', or iron bars, through both the front and back wheels of the tubs to act as brakes. But the tubs still 'slurred' on the rails, and they would have to hang onto the tubs as they went down the tunnel. At the bottom of the tunnel was the main ginney used to take the coal and the tubs to the shaft bottom. The tubs full of stone were also clipped on to the chain here.

Archer can remember his first week's wage at the pit, which was £1 2s 6d. This he 'tipped up' to his mother, and he got half a crown back. Archer worked at Bank Hall for three years, and then one Sunday night on the 'drag' on St James's Street, they heard that the local firm of Whittaker's & Clegg were doing some work down south at Croydon – so they set off down there. Archer and Leonard Redfern walked to Rawtenstall, where they slept in a hedge overnight, and then hitchhiked all the way down to Croydon, getting there the following Wednesday. They got set on at Whittaker's & Cleggs, and there were also a number of other Burnley lads there, such as Billy Green, Henry Aspinall, Wally Harrison and Joe Walsh. Archer was at Croydon about twelve months before he got called up for National Service and finished up in Palestine. Archer was in the Parachute Regiment for just short of three years. When he was demobbed in 1948 Archer came back to Burnley and did a bit of building work before going to Hapton Valley Pit. The manager at this time was Dan Warne. Fred Bullen was one of the firemen he remembered. There was also Sammy Rothwell, Harry Burns, Jimmy Holden and Ronnie Whitehead.

Archer worked on the old West Side district at Hapton Valley, which worked towards Hapton village. It was a long way in, only about 4ft high, and they had to use trams to get to the face. Down here was also a long chain road known as 'The Long Drag'. All the faces on the West Side district, numbered Two, Four, Six and Eight, were on a downward incline. This coal was later taken by the Thornybank Colliery at Hapton. Archer was put on drawing coal first at Hapton Valley on the West Side district and later did his face training with Teddy Feeney, which consisted of about sixty days working with him as a collier. Teddy Feeney retired at sixty-five, but died just twelve months later. The pits had just been nationalised about twelve months before Archer started at Hapton Valley and things were at last beginning to improve for the mineworkers. Working as a collier was hard work, but probably the most dangerous job was that of a striker, who would have to withdraw the props to be reused. They were supposed to use a device called a Sylvester, but this was time consuming and often they would just knock out the props with a hammer. The dangers in this are obvious, and many strikers got injured under rock falls.

Archer and all his brothers were decent footballers, and often played for Hapton Valley Colliery in the Hospital Cup, which is the oldest cup in footballing history, as well as in the Works League. Archer and his brothers Henry and Albert were in the team which were the Works League Champions 1952-53.

Friends made in the pit lasted a lifetime, and even now Archer is still in close contact with Mick Hargreaves, who as a youth used to draw coal for Archer in the pit all those years ago. They still meet up to have a pint and a chat. Mick was into rabbiting, and Archer into golf. While out rabbiting, Mick would collect the stray golf balls off the courses, and every now and again hand Archer a large pile of balls – he has never bought a golf ball in his life, Archer! Archer would simply say to Mick, 'I am running out of balls,' and the next day Mick would leave a pile on his dustbin in the backyard. In December 1969, Archer and his wife took over the Alma pub at Padiham, but he stayed on at Hapton Valley until July the following year while they built the pub up. 'It was only a little pub, only doing a barrel a week. Two old women had it' said Archer. His wife worked the pub during the day, and Archer would look after it at night time, after working a shift at the pit. The couple were at the Alma for three years, and later moved to the Railway pub at Padiham and kept this for another five years. Today Archer is as active as any young man I know, and he also has lots of time to spend at his caravan at Morecambe. Archer's old mate Mick Hargreaves also has a caravan at Hest Bank and the pair meet up every Saturday at a pub in Morecambe. Two other ex-Hapton Valley lads, George Knight and Steve Clough, often join them. Archer Lee and Bobby Clarke, another former miner at the pit, are also the Trustees for the Hapton Valley Disaster Fund, which keeps alive the memory of those who perished at Hapton Valley on 22 March 1962 – long may they continue to do so.

Footnote: Hapton Valley Colliery was always known as a 'family pit': you only had to mention that your father or uncle worked there and you would be set on right away. However, Archer Lee told me of an instance where this was taken a little too far. The scene was something like this: the colliery manager one day was just about to go down the pit shaft, when he noticed an old man working away oiling the tubs near the shaft top. 'Who are you?' the manager asked, 'I did not set you on'. 'No,' said the old fellow. 'My two sons work here as colliers, and they set me on.' 'We will see about that,' said the manager. 'What are their names?' Having been told, the manager set off down the pit to find the two offending colliers. 'Nay then, what's this about setting your father on – it's me who sets anybody on, not you.' 'I know,' said one of the sons, 'but if he thinks we are grafting away all day down the pit while he's sat at home doing nowt, he's another thing coming.'

∞

George Greaves
Age at interview   74
Years in mining   20
Collieries worked at   Hapton Valley, Salterford No.1, Salterford No.2, Copy, Deerplay, Bank Hall

Former miner George Greaves of Burnley gave me this interview at his home in August 2006.

GEORGE WAS BORN on Reed Street in Burnley Wood, Burnley, on 5 February 1933, the son of George Greaves (born 1910) and Alice Greaves née Paisley, and the eldest of five sons. His father, also George, started off as a brickyard man at the brickworks at what is now Rowley Tip. Later he went to the Towneley Colliery, and then to Salterford No.1 drift mine near Red

Lees and then its successor, Salterford No.2 drift, not far away. He finished up at Bank Hall Colliery until its closure in April 1971. He mainly worked on the night shift at Bank Hall.

George junior did not consider his childhood as being a rough one – he went on to say that when he was born things were beginning to improve as better medication was brought in and the sixpence-a-week Union subscriptions helped if his father was ever off work through injury or whatever. 'We did not have much, but then neither did anyone else – we just accepted this as a way of life,' said George. 'I was educated at Burnley Wood Infant School, St Stephen's School, St Peter's School, and when they could not find us anywhere else to go they sent us to Towneley Technical School'. He left there at the age of fourteen. His first job was at Finsley Mill on Finsley Gate working with his mother. After a brief spell as a driver's mate he went back to Finsley Mill till he was sixteen years of age. Then his dad said to him, 'Get yourself in pit lad', and he got set on at Hapton Valley as a pit-top tub oiler for almost two years. He then had to go to Bank Hall to do his training as a trainee miner, one week at college and one week at the pit for sixteen weeks alternately. After this he went back to Hapton Valley Colliery where he was put to work at the pit bottom near the 'creeper', turning tubs round ready to go up the pit. But at this time George was only small in stature: 'I were only four foot nowt', he said, and the tubs were coming too fast for

George Greaves outside his home in September 2006. (Jack Nadin)

him. He 'kept causing pile-ups' so the management put him at the top of the creeper where the tubs were coming a bit slower, but even then, if he came to a tub which had not been oiled he was still having difficulty – so they sent him drawing coal with the lads.

Leaving Hapton Valley, George went to work with his father at Salterford No.1 drift mine on conveyor belt maintenance. This involved cleaning under the belt beginning at the top of the drift and working his way down – when he got to the bottom he would have to start all over again – it was a continuous process. The hardest work was when a belt broke or needed extending. All the coal would have to be hand-filled off the belt before any work could begin. Then, props and wooden bars would be placed on the conveyor which was then chained up and fastened to a ring, and the motor drive of the conveyor would then be 'nudged' to take up all the tension. Once this was done, the back end of the conveyor would have to be pulled up manually to meet the other part of the conveyor belt and fastened, or refastened into place. While all this was going on, the firemen and other officials would be running around like headless chickens – because while the conveyor belt was stopped there would be no coal production. A handheld machine was used to refasten the conveyor belts, which worked rather like a large stapler, putting 'staples' across each of the narrow sides of the belt, and then a wire rod was pushed through the 'staples' to make the fastening connection. George told me that his father lost half his thumb doing this sort of work.

By this time George was doing a bit of 'courting' and thought that it was time he moved on to better money by working on the actual coalface. He began work on the coalface by 'scufting' or cleaning up the small coal behind the coal cutter – he did this work for about three years. George was now about to get wed, so he upgraded himself once more by going for coal-cutting machine training. About this time he became acquainted with Albert Woodward and Fred Lockier, who became his training officers. He got his 'papers' for the shortwall cutter, and later the longwall cutter, and when the Anderton Disk shearers came in, he got his papers for those. About this time he moved on to Salterford No.2 pit, and then to Copy Colliery at Cliviger. It was at Copy that George was put in charge of his first coal-cutting machine – Mr Malone was the manager at this time, he was also in charge of Salterford Nos 1 and 2. The Copy Colliery was closed down in 1964, and by March that year all the salvage work had been done. Some of the workforce went to Bank Hall, and others – including George – went to the Deerplay Colliery behind the Deerplay Inn on the Bacup Road. Deerplay was another drift mine where you could walk into the pit, but ride out on the conveyor belt. While George was working at Deerplay Pit two lads got killed – he remembered having to carry one of them out of the pit.

At Deerplay George was on 'dattaling', or day work, doing things like striking props and backening for the rippers. Other times he was put on supplies, getting props and timbers to the coalface. When Deerplay Colliery closed down in 1968, George was moved on again, this time back to Bank Hall Colliery at Burnley. George never liked Bank Hall Pit because of the shafts there – he was all right in the drift mines but had a phobia about shafts ever since his father was plunged into the sump at Towneley Pit and nearly drowned. It appears that the winder at Towneley had a heart attack or something and the cage crashed into the water-filled sump. Because George senior was also short in stature, two of the taller men who were also in the cage had to lift him up to keep him out of the water otherwise he would have drowned. The winders at Bank Hall did not help; many of the other miners had worries too about going down. The winder would raise the cage a little at the top, then drop it like a stone down the pit, 1,500ft to the shaft bottom. Those at the bottom would not escape either; they would be 'yanked' up and away from the pit bottom. I think it must have been a stomach-churning experience for all who worked there – no one could say that they ever got used to it! In the end, it all became too much for George and he packed it in. This would have been around late 1969, just eighteen months before Bank Hall was closed down forever.

Following this George went on to do various jobs, including driving work, having got an HGV licence, and working on the scrap metal and demolition work until he retired in August 2005. George and his third wife Barbara, whom he married in 2002, now enjoy their caravan at Morecambe and go over there at every available opportunity.

Steve Hird
Age at interview   61
Years in mining   15
Collieries worked at   *Bank Hall Colliery, Burnley, Hapton Valley Colliery, Burnley, and Yorkshire Collieries on salvage work*

Pickup Croft might be remembered by older residents of Burnley. The site today is that roughly taken by the bus station off what are now Centenary Way and Croft Street. It

Steve Hird at the National Coalmining Museum (Caphouse Colliery) near Wakefield. Steve still has an interest in anything to do with mining. (Steve Hird)

was quite a little close-knit community consisting of streets such as Peter Street, Norton Street, Miller Street, Pickup Street and Hatter Street.

STEVE HIRD WAS born on 22 October 1946 at Bank Hall Maternity Hospital, Burnley, the eldest son of John and Madeline Hird. Madeline's maiden name was Parker and she originally came from West Byfleet in Surrey. The other children were brothers Michael and Bernard and sister Judith. The Hird family were connected with the Pickup Croft area for generations – from 1860 through to 1944.

Steve's grandfather William Henry Hird was at Passchendaele, the Belgian village which marked the furthest point in the British advance during the Ypres offensive of 1917. Eight weeks of heavy rain made the conditions for the exhausted armies almost impossible and there was heavy loss of life. William Henry was gassed during the war, and in his later life he served as a 'bookie' when betting was still illegal in Pickup Croft. Despite this, some of the more prominent gentlemen in town placed bets with him, including Dr Purves. Steve's father John worked most of his life at the Burnley Brickworks, near Bank Hall, apart from when he served as a regular soldier from 1938 to 1946; he too saw action at both Dunkirk and Gibraltar. His mother Madeline came to Burnley at the end of the war and married Steve's father – they had met at an army dance in Woodall Spa. Childhood days were not easy for many just after the Second World War when Steve was born, but Madeline contributed to the household budget by making clothes and doing a bit of needlework. 'There was always something wholesome on the table when we arrived home from school though', added Steven. The family left Pickup Croft in 1946 to live in Lindsay Street, where they stayed until 1960 before moving to Rosehill Road. All the boys, including Steve, attended St Mary's Roman Catholic School under the 'Culvert' and then St John's School in Ivy Street, finishing at St Mary's Boys' School. Their sister Judith attended St Hilda's School. All the children left school at the age of fifteen. Steve has some happy memories of Pickup Croft and St Mary's School. Teachers he remembers at St Mary's included Richard Kelly and Jack Cavanagh.

Many happy hours were spent along the 'Straight Mile' canal embankment on the Leeds & Liverpool Canal. Near here was the 'Destructors Yard' where abandoned and or unwanted pets were taken to be put down. The kids used to hate thinking about the animals being put to sleep. One day they were watching a man carrying a cat by the scruff of the neck to be put down. A carefully aimed stone thrown by Steve hit him on the back of the head and he dropped the cat. The animal wasted little time in hanging about and headed straight into the long grass on the embankment to freedom. They 'avoided' the area for the next few days in case they got found out.

On another occasion around Bonfire Night Steve dressed up a lad he knew as a Guy Fawkes, stuffing him with newspapers and sitting him down outside the Odeon cinema. The prank worked extremely well and they made about £2 in cash from the unwitting passers-by. However, a large crowd had gathered to view the 'Guy', who had to keep perfectly still while all this was going on otherwise he would have been found out, but by this time he wanted to go to the toilet – and as the crowd increased he wet himself. His mother was not pleased when he got home and he got a good leathering, as they say in Lancashire.

There was also a 'clogger' on Pickup Croft who some might remember – his name was Matt Shaw. A devout Catholic, during the 1920s and 1930s he would contribute clogs to those children without footwear, whether they were Catholic or not.

Steve worked in the building trade at first, 'on the trowel', and started in coalmining in 1969/70, doing his basic training at the Bank Hall Colliery, the wages being better than in the building trade. He then went to the Hapton Valley Colliery, starting off as a 'tackle lad'. Both his face and ripping training were done under Jimmy McCarthy and Bernard Bispham. It was Jimmy McCarthy who showed Steve how to use the Eimco bucket loader, a compressed air shovel rather like a small bulldozer. This was used to load the ripping rubble onto the scraper conveyor, the larger rubble being kept back for building the pack walls at the tunnel sides. Len Mack was stationed in the pack hole. Being a 'Brummie' he was always good for a laugh. Steve could remember the pit bottom at Hapton Valley: you got off the manrider at the mouthing, the manrider guard usually being 'Red' Jack Oldham. If you were working up M6 or the M8 Districts you had to go through the air doors to an endless rope manrider to the 'Scouring', where you had to walk the last 200–300 yards. If you were working in the 'Dip' you stayed on the manrider and turned off left past the bunker, where a lad called Milton worked – Milton would always give you a toffee. The lads who worked down there were John Irvine, David Latimer and Tony Mitchell. Both John and David had been respectable stonemasons and joiners. David, 'Lat' as he was sometimes known, worked in the joiners' shop. Part of the management team at Hapton Valley were Bill Ferris, manager from Whitehaven in Cumbria, and Harry Cooper from the North East. Colliery officials were Bob O'Hara, who Steve still keeps in touch with, a fair but firm man, and Tommy Chapman, under manager, who sadly is now deceased. Another who has now passed away was fireman Billy Ellins; he once reprimanded Steve for coming on to his district without reporting to him after he had put a lacing in the conveyor belt. There was another fireman named George Holden – he came from Accrington.

There were many the stories told, some of them quite comical now, like the time 'Red' Jack Oldham and Bob Gregory were at the Miners' Home at Blackpool. One day they were both sat outside the home on benches watching the world go by, when one of the Blackpool trams jumped off the rails. Jack and Bob watched the scene for a while, and eventually a couple of fitters came along armed with a Turfor (a pulling device) and chains to try and get the tram back on the tracks. It was soon obvious to the pair of miners sitting there that the two fitters had little idea of what they were doing, so Jack and Bob went to their aid. They lashed the chains to a nearby bollard and pulled the tram back onto the tracks. All this of course took place while the men were meant to be convalescing from the rigours of pit work. This is a story which has now been passed down through generations of miners – almost a part of the annals of mining folklore. Nevertheless it is true. 'Red' Jack was often seen on his way home from Turf Moor, Burnley's football ground, but would never tell anyone the score. If asked he would say, 'Thi can go to Turf Moor thee self and pay your money like I had to do'. The drift top foreman, Jimmy Barrett, would always be careful what he said about the 'Clarets', especially if they had been beat, to 'Red' Jack, because Jack would inevitably remind him of an earlier boxing match in the Inter-Colliery Sports, when Jimmy was favourite to win, but lost the match.

Billy 'Brad' (Bradbury) would put it around that Steve was 'Red' Jack's son, because they both had red hair. Another man Steve remembered was Dick Bentley, the branch secretary of the Union at Hapton Valley – he would do anything for you, going out of his way if need be. Dick has now passed away, but he will be long remembered. Walter Thorpe was another man who would always be willing to help. When Steve's younger brother Bernard died suddenly in 1978, Walter was the first, but not the last, to ask if there was anything he could do to help. Walter has also passed away now – but it is memories of friends like these that you never forget. Take yet another man, Harry Cooper the under manager; because of Steve's bereavement he had to take some time off work, and Harry gave Steve a week off work on day wage.

Then there were the 'characters' at the pit. Bobby Clark – he would liven up a pub in no time. Harold Durant, David 'Brains' Pearson, Teddy Feeney, also known as 'Rocking Teddy', Billy Moran and Tony Cheetham, who had the rather strange habit of jumping out of manholes down the pit pretending to be Quasimodo – he was a real practical joker. Steve also recalled John

Pinder, who died recently, another great man. John also enjoyed his day out at Turf Moor supporting his home team in the 1970s. The 'Clarets' had a goalkeeper at this time, Alan Stevenson, who John once saw let in a soft goal from the 'Bee Hole End' of Turf Moor. John never forgave him, and often criticised his every move from that game on. Stevenson got used to John's remarks, and even acknowledged them as he ran to the Bee Hole End goal. It was all good-natured of course – all part of the game. Then there was Peter 'Percy' Pounder and John 'I insist' Hughes – the list goes on.

Steve worked at Hapton Valley as a beltman on the afternoon shift with Eddie Bright. They had an arrangement that if Burnley had a night match, then Steve would come up the surface drift in the pretence of checking the surface belts in the screens so that he could get away early for the match. Occasionally the afternoon foreman Alan Gregg would find out, and threaten to stop his pay, but never did. 'They say the lads you worked with in the pit are the best you will ever work with – and it is true', said Steve, who worked at the pit until it closed down in 1982. He now works as a coach driver and tour manager for a well-known firm in the English Lakes, but has never forgotton his days down the pit, nor the lads he worked with – they were the salt of the earth. Given the chance he would do it all over again, and still have no regrets.

Stuart J. Ingham
Age at interview   60
Years in mining   20
Colliery worked at   *Huncoat, Huncoat near Accrington,*
                           Thorneybank, Hapton near Burnley, Hapton
                           Valley, Burnley

Former miner Stuart Ingham now lives in the village of Belthorn near Blackburn – he was a fixed-plant fitter at the collieries where he worked. Stuart was able to pass on some of his memories of his coalmining days to us after he had contacted me by letter. Stuart still has a wicked sense of humour as we will see from his recollections of those days – and was also able to tell me something which I always believed in, and that is that the Hapton Valley Colliery was haunted. Here Stuart takes up his tales of coalmining in East Lancashire.

I WAS BORN in Rossendale General Hospital in 1946, and brought up in Accrington. Early life was basic, as it was for most people then, a two-up two-

Stuart Ingham at his home in Belthorn village in mid-November 2006. (Jack Nadin)

down terraced house with a toilet at the bottom of the backyard and a coal hole. When it came to starting work, Mother said to me, 'I don't care where you go to work, but you are not going down the pit'. She had relatives at the pit, and most of them had a bad time of it. Most of my mates at that time were starting apprenticeships as joiners and plumbers and were working for something like £2 a week. When I saw this job advertised for trainee miners at Bank Hall Colliery who could earn £5 16s a week, I went for it. My father was only on £7 a week then.

The person I saw was Ivor Jeremiah, the main training officer at Bank Hall at that time. After training I was put on as a tackle lad and other jobs, but decided to put in for an apprenticeship. The National Coal Board did not just want young lads as miners, they also wanted electricians and mechanics and actively encouraged those who wanted a trade in the industry. I was posted over to Huncoat Colliery along with a few of the other lads I had become mates with, one in particular I recall was Vincent Higham. However, we never went down the pit at Huncoat, I finished up in the loco shed. They had an old guy there, and I was put under him. I am not sure whether he was called Bill or Bob, but what a powerful old fellow he was. As far as the management were concerned this fellow was like a god, because he was the only one who could mend the trains. Huncoat Pit had extensive railway sidings, and if they were not able to move the coal on the surface, then everything else at the pit came to a halt.

We had a brew shed on the pit top, and I was with this Bill or Bob, whatever they called him, one day when the manager and some other officials came in. Bill, or Bob, shouted out to them: 'Get thee self on the other side o' that door, and knock on it, or else tha not coming in.' And the manager and his team obeyed! But even then he said to me, 'Count to ten and then go and let them in'. That's the sort of respect this fellow had at the pit – what he did not know about trains was not worth knowing. He never liked to pass on his skills to anyone else, but he did to me. I don't think he thought of me as being a threat – even so it was me who got all the dirty jobs such as lighting fireboxes and cleaning out the grates. Unofficially he used to let me take the trains out, down as far as the main line. I think one was called 'Raven' and the other 'Linnet'. You can imagine how proud I felt stood on the footplate running these trains up and down, I was still a young lad at this time, and this was every young lad's dream. One day though I nearly tipped the train over. I went out in a snow blizzard, and as I approached a sharp bend the train leaned over – I am sure it went round on two wheels. That frightened me to death – that really shook me, because these trains are big things, not like a car. After this it taught me a lesson, and I had more respect for the trains. After a time at Huncoat I was sent to Thorneybank Pit, a new pit just down the road from Huncoat, near Hapton village. Here I was put with a man called Jimmy Appleby; he used to be a professional footballer for Burnley FC. He was a Geordie lad, but when he retired he liked it here so much that he bought a house down here. I liked it at Thorneybank, the manager at that time was a chap called Jim Yuill – he was a good manager. I used to go shooting with his son, also called Jim; we would go to competitions and so on.

I was on a coalface at Thorneybank one time, with what they called a white lamp: where normal cap lamps are black these were white. You wore them if you were not face trained. They put me on a little coalface which just had an AB coal cutter on it, and the safety officer spotted me. He shouted me down and asked me if I had done my face training, which of course I had not, and then I got a right blasting off him. It did not make much difference though, they just gave me a black lamp instead of a white one.

I was once coming out of the drift mouthing at Thorneybank and near here was a cabin with pumps in it – being mechanically minded I had a look in. All the joints on the pumps were leaking, which I know now they should have been, because they acted like a coolant for the pumps. Anyway, I did not know this then, so I tightened them all up and stopped the leaks. Later I saw one of the bosses in the canteen, and said 'I have just done your lot a big favour' and told him what I had done. Well, he went barmy, called me all the names under the sun and ran over to the cabin which was just about to go up in flames. All part of life's sweet tapestry I suppose, another lesson learned.

When the explosion happened at Hapton Valley I was walking home down Empress Street at Accrington and noticed all the womenfolk outside their houses talking about it. By the time I had got to my house my mother grabbed me – she had been worried sick. They had heard about the explosion on the wireless, the early reports saying that the explosion had happened at a pit in Hapton, and of course Thorneybank was at Hapton. I jumped on the next bus, and instead of getting off at Thorneybank I carried on to Hapton Valley to see if I could be of any help. It was absolute chaos there. I did help out, but we were not allowed to go underground because you had to have a card to get down the pit. There certainly was no lack of volunteers, and what amazed me was the large number of families there, relatives waiting for news, and so quickly after it happened.

A short time later I got moved on to Hapton Valley. I can still remember the crosses chalked on the rings where they had found people injured or killed in the explosion. They had the Inquiry of course afterwards and what they put it down to was a spark from the foil used to wrap up the chewing tobacco. They reckoned that if this got compressed and compressed over again, it became tight enough to make a spark and ignite the firedamp. After this the tobacco used to come in yellow and brown greased paper. But what a shame about all them lads eh? The manager then was Adam Weir. By this time I was getting interested in the Miners' Union, and became a Union representative and later a delegate for the men. Other Union men I remember were John Barker, Dick Bentley – he was the Union secretary, he was a good man – and Mick Hargreaves.

I can say one thing about the people who worked at Hapton Valley: they were a great set of lads, a good set of workmates, I could not fault any of them at all – they were absolute gems, every one of them. Some of the under managers I can remember were George Turfrey, I could never get on with George. Someone told me he had a bad heart, so I used to hide on the manholes when he was coming up and jump out at him. Another under manager was Geoff Blackburn, he were a bugger, I got him as well. He used to crawl through the coalface sweating because he was a hefty man, and every so often he would stop and take a swill out of a water bottle, it did not matter whose water bottle it was, he would just help himself. He never brought his own water down – so one day I had a pee in my bottle and put it to one side and sat down to wait. Sure enough along came Geoff and took a big swig, he must have thought that it were pop. They played horrible tricks down the pit: they would put dead mice on your butties or in your water bottle. Tackle lads were the worst; they would spread a bit of crap on your butties, or smear some around the telephone mouthpiece. The person using the phone would say there is a bloody stink around here, little did they know it was right under their noses!

I got my leg crushed one time up No.12 face when the entire gob came in. The gob had been hanging for about three days; we kept shouting for them to get some timbers under it, but they never did. On the Wednesday night when I had gone in on this job, I had finished the job when all at once all the lot came down and crushed my leg. That basically was the end of my coalface work. I was married then and off work going for therapy trying to get my leg going again. When I did go back, I was asked if I wanted to stay on the pit top on light duties, but I said I would go back as I was on nights. I think I had an idea the first night when I went on the face that it was not going to work. By the time Wednesday had come around, I was going up the face to have a look at a shearer which had broken down, when Hebert Boys the overman asked what was wrong with me. I said I will be OK, but he insisted on having a look at my leg, which by this time had all swelled up and was a right mess. Hebert said let's get you down the face and out of the pit – but he insisted on getting the shearer going before I left. After this the doctor told me that I was not going to be able to work on the coalface in the future. So, after this I was put on working on things like the manriders and the main bunker underground.

We used to have a few ploys when it came to management or other underground officials, especially if there was a breakdown. They would ask how long we were going to be before it was repaired, and I would say three-and-a-half hours or thereabouts – then their bloody heads would fall off. Of course when I got it going in half an hour I was the best fitter in the pit.

I worked on the manriders with a lad called Geoff Green, sadly he has died.

'What about the ghosts?' I asked Stuart.

Well that was 'Charlie'. How we knew it was him is because he always wore one of the old-type helmets, the old compressed paper ones, and a dog collar with a ring on the front. He had a red face and also wore clogs. One day a tackle lad was working next to the crab at the top of the manrider road and he heard someone coming down. 'Charlie' stopped and asked the tackle lad how long he had been down the pit, to which he replied, 'Just a few weeks'.

'Charlie' said to him, 'Well just do as you are told lad and you will be alright,' and then he continued on his way down the manrider road. A little while later the manrider came up. The guard was Bob Gregg, and the tackle lad said, 'Who the hell was that you just passed down road?' Bob said, 'Nobody's gone down there, I would have seen him'. Bob asked the tackle lad to describe the man, and when he did, he went all cold. Well, that's Stuart's version of the ghost at Hapton Valley Pit, anyway – or might this perhaps be just another of Stuart's pranks? On some occasions we might have to go to other pits in the area. One time we were posted up to Hill Top Colliery at Bacup – that were a funny little pit. We used to get sheep coming into the mechanics' shop there to warm themselves on the potbellied stove we had. Tommy Smith was the main fitter up there at that time. Eventually he came to Hapton Valley. When they were commissioning a new coalface it would be Tommy who sorted it all out. We also did a bit at Scaitcliffe Pit, but not down the pit itself. Jimmy Appleby, the fitter at Thorneybank, was there with me. We had to drop one of the cages at Scaitcliffe after the pit had finished, we burned through it and let it drop down the shaft. Jimmy was also with me at Calder Colliery. This had been on fire and was all shut off then – we worked getting stuff off the pit top and round the baths there which were moved on to other pits such as Thorneybank. The pit had been on fire and they had to seal it off.

Stuart was a real prankster and practical joker, as he himself relates as he continued his tales of coalmining.

I was working in the workshops on the pit top one day when this representative came along and asked for Eric Holland, the main fitter. I told him that he was down in the compressor house near the top of the drift. I pointed out where this was, and asked the man if he had met Eric Holland before. 'No,' he said. 'Well, he is a bit deaf, in fact he is stone deaf,' I told him. 'What you will have to do is shout out loud, rather than just talking to him – he should be able to hear you then,' I added, and sent him on his way. Meanwhile I got on the phone to Eric and told about this man coming over to see him, and told him that he seemed to be a bit hard of hearing, and that he would have to shout out to him. I never actually saw it, but it appears that the two men were shouting at each other and making half-hearted gestures with their hands for about ten minutes. It was then that one of the other fitters poked his head through into the compressor house and said to Eric there had been a breakdown in the pit. 'OK,' said Eric, 'I will look into it'. The representative then said, 'I thought you were deaf,' and Eric said, 'I thought it was you who was deaf'. It was then that they both realised that they had been taken in.

When Hapton Valley finished, there was a scheme going round whereby you either got redundancy money or another job. Sid Vincent and Joe Gormley, the main NUM men, got involved. Eventually we came up with our own scheme; we would take the redundancy money, then sign on the dole. About three weeks later we would then apply for another job at another pit, in effect we managed to get both the halfpenny and the cake. A few of the lads we left behind at Hapton Valley to salvage the pit, such as Raymond McGoogan and Brian Maden. I once went back up to the pit, after it had finished like, to take some photos, but they would not let me. Some roughnecks came down and shouted, making it clear I was not allowed on the site. I tried to explain that I used to work there, but they cleared me off saying that I was trespassing. I got this job at the Walkden workshops near Manchester thinking I would be servicing chocks and things like that. At the interview I was asked by one of the men if I knew anything about loco engines and I told that I used to work with them at Huncoat Pit. 'Right,' he said, 'You can come and work with me'.

Walkden at this time was taking in contracts from all over; they even had the 'Flying Scotsman' in. I was shown into a large workshop where the noise was deafening, all these men riveting with pneumatic guns – you could not hear yourself speak it was that loud, and not one had any ear defenders to protect their hearing while all this was going on. Later I heard about a fellow at Accrington, a watchmaker who was selling up. After a bit of bartering I bought the shop off him even though I knew nothing about watchmaking. So I got some books on the subject and began to practice on some of the watches in the stock.

The first one sprung apart, so that went into the bin. I got the second one back together, but it did not work, so that went into the bin, but eventually I managed it. In time I was taking all the trade from the other jewellers in town. I was also selling fancy goods, but even so there was not much money in it. In the end I managed to get a job with the Water Board. I tried to run the shop as well for a time, but that did not work out. So I went for the Water Board full time, a job I have had ever since. For the last twelve months though I have been off work as I have been having bother with my legs. Although I am on full pay I am trying to get back to work, I have got the Union on it. I enjoyed my time at the pit though – my happiest times I think were at Hapton Valley – they were a grand lot of lads there. I would go back tomorrow given the chance.

Hoddlesden Colliery

The Hoddlesden collieries were begun by Joseph Place around 1838 and consisted of a number of pits worked mainly for fireclay, although coal was mined at a later date. The last of these pits was the Hoddlesden No.12 Colliery dating from 1934 and located on Hoddlesden Moss, above Darwen, consequently it was always known locally as 'Moss Pit'.

The pit survived into nationalisation and pithead baths were erected by the NCB in 1953. The extraction of fireclay ceased in 1952 through lack of demand and production concentrated on getting the coal. On the surface there was a pumping shaft and surface drift with screens, pithead baths and a canteen. When the colliery was closed down on 29 September 1961 it employed ninety-nine men and had produced in its last full year of production 17,743 tons of coal.

∞

Richard John (Jack) Counsell
Age at interview   67
Years in mining   7
Collieries worked at   Bank Hall, Burnley; Hoddlesden, near Darwen

One man among the ninety-eight others who was there when the pit closed down was Richard John Counsell, who was kind enough to relate his memories of the 'Moss Pit'.

I WAS BORN on 30 September 1938. My earliest recollections of life begin when I was about six years old, being brought up with my three brothers, William (Billy), Roy and Derek (Dek), and two sisters, Rita (Reet) and

The new bunkers and gantry were erected at the Hoddlesden Colliery in 1957. The colliery was closed down in September 1961 and the last year's full output totalled 17,743 tons, when the pit employed ninety-nine men. (W. Rawstron Collection)

Mildred (Milly). We lived at 21 Great Bolton Street, Blackburn, a terraced house, with my Aunty Frances (Fanny) and my dad, John, and mum, Mary Gertrude (Gertie). The house was on a main road and consisted of two bedrooms, a living room, and a parlour, which you entered off a lobby, and a kitchen. There was no hot water and the toilet was in the backyard and it always froze up in winter. In the living room was a cast-iron fireplace. It had a water boiler, which never worked, on the left-hand side, the fire grate was in the middle and it had an oven on the right-hand side, which was heated by the fire. Above the fire, attached to the ceiling, was a clothes rack worked by a pulley system where washing was hung to dry out over the fire.

We all followed in one another's footsteps as far as schooling was concerned, all going to the Park Road Primary School, Blackburn, and then we all went to Audley Secondary Modern School, Blackburn. Times were hard, as it was just after the war, and there was no work to be had no matter what the government had promised. I remember being in the house one day when my dad was leaning on the mantelpiece – those old mantelpieces were about 5ft tall – and he looked really depressed. Here was a man who had just spent five years fighting for his country, and he was being denied the basic right to work and to support his family. That night there was a knock at the door. Fanny opened the door and a big rugged man came striding in. He looked at my dad and said, 'I heard you are looking for work'. My dad replied, 'You heard right.' The big man asked him if he had ever laid drains. 'Plenty', Dad replied, even though he had never laid a drain in his life. The big man said, 'Right, start tomorrow.'

John Counsell outside the pithead baths at the Hoddlesden Colliery. (John Counsell)

Four weeks passed and the big man called again. 'Well John, just as I thought you have never laid a drain in your life have you?' Dad looked at the big man and smiled, saying: 'I am sorry, but we were desperate for the money and at least I did all the digging.' The big man laughed, 'I know and I hope it helped you out a little, but I will have to let you go now, I hope you understand.' Dad said he did, and added: 'Thanks a lot, I have got a job now at Clayton & Goodfellows foundry – I start on Monday.'

Things seemed to get a little better from then on, as Mum also got a job at Goodfellows. Billy was ready for leaving school – you left when you were fourteen then – and Derek and the girls were growing up. There were a lot of pubs in our neighbourhood. Mum and Dad's local was the Nelson on the

Richard John (Jack) Counsell in his garden in September 2006. (Richard John Counsell)

corner of Nelson Street and Park Road. We lived just two minutes away and Dad went in most nights; he earned the right because he worked hard. He would come back from the pub about eight o'clock each Wednesday and Saturday to take Mum out for a drink. We were looked after by our Fanny; we never called her aunty. Although we were poor we had a very happy childhood; the warm summer days and nights seemed to go on forever. We would play out until it got dark, then you would hear our Fanny calling us all to go home: 'Jacky, Roy, Mildred, Rita, Derek, come in now, bedtime.' Ah, those long, warm summer nights, with nothing to be afraid of except a clout from the policeman if you did something wrong – you didn't do it again!

Bath time! That consisted of a tin bath usually hung on a peg in the backyard; it was brought in and placed on the living room floor in front of a coal fire, and filled with water which had been boiled in a kettle. Then, at bedtime, it was Roy and I in a double bed with Billy in a single bed in the front bedroom, Derek, the girls and Fanny in the back bedroom, and Mum and Dad in the front parlour. Buckets were under the bed for peeing in on winter nights; remember the toilet was in the backyard. The bedcovers consisted of army blankets and some old coats.

In those far-off days, no one really had money to spare, and most grocers would let you get food through the week and pay them on payday: this was called 'strapping up'. Our shop was at the corner of Canning Street and Nelson Street, Blackburn. Sometimes you couldn't pay it all on payday so you paid what you could and 'straightened' it up when you could afford it.

The pupils in my class at Park Road School, Blackburn, were about to take their exams to see if they could they win a scholarship to either the technical college or the grammar school. This exam was taken in two parts. The first part was taken at your own school, and if you passed that you took the second part at Blakey Moor School, which was close to the police station in the centre of Blackburn in an area which was called Blakey Moor. Seven of us passed the first part, six girls and myself. Out of those seven only one passed the second part and that was me. When my mum and dad and Fanny heard they were really pleased for me. But reality set in when I was given a list of the things I had to have, including new clothes, sports equipment etc. So I ended up at Audley Council with most of my mates anyway. I was picked to play for the school cricket team as a fast bowler, and I even got my name in the local paper a few times, the *Blackburn Times*. I also threw the discus for Blackburn, and was champion of Blackburn Schoolboys and around 1953 won a medal. I joined the Boys' Brigade, but preferred to go fishing on Sunday when I was supposed to go to Sunday school. The Brigade leader told me that if I missed two Sundays together

I would have to quit the Brigade. So I went to Sunday school one Sunday, and went fishing the next Sunday, alternately. When it came the time to leave school, my dad asked me where I would like to work. I thought about it for a while then I made my decision: 'I want to work down the pit.' 'Are you sure?' asked Dad, and I told him I was positive. So I finished school on Friday and started work down Bank Hall Colliery at Burnley on the following Monday.

You had to do sixteen weeks' training, eight on the surface and eight underground, alternating weeks, and I was the only one from Blackburn, but there were a couple of lads from Darwen, Joe Walsh and Bernard Brooks (nicknamed Bunt). These two were already well into their training, and had only a couple of weeks to go before they went to the coal mine of their choice, which was Hoddlesden Moss, near Darwen. There were forty of us at the start but for various reasons a few dropped out.

At Bank Hall Colliery the trainees were not allowed in the baths, even if you had been working underground, so you went home in your dirty clothes and with your pit hat on. The bus to Blackburn was always a Ribble 'White Lady' and it was immaculate; the driver even once offered us money not to get on.

While I was working on the surface at Bank Hall I became involved with alcohol. A few of the lads said that they were going to this certain shop, which might have been called Fitzpatrick's, at dinnertime, and would I like to go with them. I said that I would, so off we went. When we got to the shop they introduced me to 'hop bitter' made from spent hops brewed again. There we were, just fifteen years old, and drinking bitter.

When I arrived at Blackburn from the pit at Burnley I would have to walk the rest of the way home. Everyone else was either at work or at school, so I would have to climb over the backyard wall and go in through the backyard door, which was always unlocked. One week when I had been training underground I went home as usual and, as I was climbing over the backyard wall, I heard a scream. I rushed in and there stood my mum. 'You little devil, you frightened the life out of me, and I wondered who it was climbing over the wall,' she said. She had never seen me covered in coal dust before.

The weeks passed on, and I was coming to the end of my training. Joe and Bunt had already finished a few weeks earlier and they had started at the 'Moss Pit'. I had also chosen the 'Moss' as my choice. I picked up my last wage from Bank Hall Colliery; it was £2 10s.

I started work at the Moss Pit on the following Monday. The Hoddlesden Moss Pit was a drift mine. There was a shaft but it was only used in emergencies, and the cage there only held four people. The manager was Richard (Dick)

King, and the under manager was Alexander (Sandy) Mitchell, the only man I have never heard swear. The mine was about a mile from Hoddlesden village, and really consisted of two pits in one. The Top Mine (a seam is always a mine in Lancashire) produced the coal and the Bottom Mine produced clay – I worked in the Top Mine. When the coal reached the surface it was loaded onto ten-wheeled NCB wagons and dispatched. When the clay reached the surface it was sent by tub on rails over the moor to Hoddlesden to a gang waiting at the end of the rails and from there it was sent to Places Pipe Works, where they made drain pipes, tiles etc. The place at the end of the rails was nicknamed Bert's Brew simply because the man in charge, who worked for the NCB, was called Bert. To get to the pit a special double-decker bus was laid on just for the pit lads. We caught the bus at the bus station, which was at the centre of Darwen at a place called 'The Circus'. This went over the Grane Road and turned off at the Grey Mare pub onto Jackson Heights and onto the pit top. Once off the bus you went into the canteen for a quick brew and then into the locker rooms to get changed into your pit clothes. You then went into the lamp room, picked up your cap lamp and your check. This was a brass check with your personal number on it; this was a safety thing (my number was 88). If there was an accident or something, by checking the board they would know with the number of checks missing how many men were down the mine. In a serious accident perhaps the only

The entrance gates to the Bank Hall Colliery on Colne Road, Burnley – passed by thousands each day from the 1950s until the pit was closed down in 1971. The gates still survive at least in part, being the entrance to Bank Hall Park, the large open space which replaced the coal mine. (W. Rawstron Collection)

means of identification would be your check. When you came back up the pit you put your check back on the board and they would then know you were out of the pit.

When it was winter the bus often could not get over the Grane Road, so we would get to Hoddlesden, the roads having been cleared by the gritters, and get a lift over the moor to the pit by riding on the tubs from Bert's Brew. The pit worked a system called the pillar and stall method, no longwall or shortwall faces. A Siskol coal cutter used to cut the coal, and there was never any conveyor belts at Moss Pit. You worked with two colliers; they filled tubs with coal and you drew or pushed the tubs on rails to what was called the wheel end, a job known as drawing. When you reached the end of the rails you were at the 'wheel end' in a higher area of the workings. On the ground here were big steel plates known as shunts, which made it easier to manoeuvre the tubs and put them on the main ginney to be taken out of the pit.

The tubs were sent to the bottom of the drift by means of an endless chain haulage system, known as a ginney, which was worked by an electric motor. The full tubs went down one side and the empty tubs came up the other side on two sets of rails. While I was drawing I took a full tub out from the coalface and took an empty one back. Each gang had two coalfaces to work. While one was being cleared the other one was being 'cut'.

The Siskol cutter had to be level or it would not work properly. It cut the coal about 6in from the bottom; if the cutter was set too much at an angle it would catch the floor half-way through the cut and you would have to reset the cutter. On top of the Siskol cutter was a big ratchet with a steel plate on the top; this you wound close to the roof, where a long wooden bar with a prop under each end had been placed, then you wound the ratchet up to this and tightened it up, securing it to the roof and floor. You had to keep tightening the ratchet as you cut the coal. On the side of the cutter were two large handles, one for rotating the cutter, the other for winding the rod closer to the coal.

The cutting involved six rods: when the first rod had gone as far as it could you changed it to the second rod and so on, until you had used all the rods. Then you moved the cutter and drilled the holes in the coal as near to the roof as possible. You drilled about four holes. You then called for the shotfirer to stem it, load it and 'fire' it. You got down onto the pit by riding on the bogies; these consisted of a set of three flat wagons on rails linked together, and each wagon would accommodate six men. This was then lowered down the drift by a steel rope. While it was being sent down the pit empty bogies came up the other side. When the empties reached the surface they were physically moved to what was called the loading side. When you reached the bottom of the drift, which was 365 yards deep (as many yards as days in a year), the miners who worked in the Bottom Mine walked down. I believe you could stand upright in the Bottom Mine, although I never worked down there. The men who worked in the Top Mine got on a small flat wagon, known as a tram, and you pushed yourself with one leg along to the coalface using the sleepers on the rails.

All the main roads to the coalface were barred and propped. Moss Pit was a semi-dry pit, although some of the faces were wet, for which you received 'wet money', but the others were dry. The roadways were on average around 36in high. When you wanted the rails extending on the roads that went up to each face, you sent for the maintenance man, he was called the dataller. When you returned to the surface you went in the lamp room, put your check on the board and put your cap lamp on charge. I was first put to work with two colliers named John Morris and Jimmy Brighty. John was a small bloke and Jimmy was a stockily-built fellow. I had never done any drawing before so they had to show me how it was done. When they filled the tubs, I got behind it, and, because it was on an incline, when I put my feet on the rails the weight of the tub pulled it forward, dragging me behind sliding on the rails. If you got the arch of your back too high you would hit the roof bars and graze your back. By the time you had taken thirty tubs out and brought the same amount back you were in some considerable pain.

Then when you came out of the pit you went to the showers, and this was my first experience of getting bathed with lots of other men. The baths consisted of about fifteen showers (no cubicles) and you asked one of the other colliers to wash your back, you in turn would then wash his. Then one man said to me, 'Come here son and I will wash your back'. I know it is a long while ago but I will never forget his name, it was Bob Eccles. He had a few nephews working at the pit, as well as a brother. I should have known something was wrong when John Morris said, 'Be careful Bob'. I did not know at the time that the colliers used washing powder for cleaning themselves, and this is what he started to wash my raw back with. I didn't give him the satisfaction of seeing me wince though. When you came up from the pit into the baths you took your dirty clothes off and put them in the 'dirty' locker, and when you had bathed yourself you went to the 'clean' locker and put on your clean clothes ready for going home. The lockers were heated, so that when you came back the next day all your clothes would be dry and ready for you to put on for work. If you had time, you went in the canteen for a quick brew before you got on the special bus, which took you back to Darwen Circus.

The bath attendant at Moss Pit was a man called Jack Wild; he had two sons who worked down the pit, Harry and Charlie, both of whom were shotfirers. Jack was a small man, and Harry and Charlie were small in stature too, but both were powerfully built. Jack also had another 'occupation': he was a bookies' runner. Betting shops, or bookies, were illegal at that time.

It was hard work at the Moss Pit, but I really enjoyed it. I had obviously met up with Joe Walsh and Bernard (Bunt) Brooks again, the lads I did my training with at Bank Hall Pit, and I used to go over Darwen at the weekends and go for a drink with them. If I was a bit worse for wear and could not get home I stayed at Joe's house; he lived with his mum and dad on Maria Street in Darwen. The Moss Pit had been nationalised about ten years when I started there in 1953. When it was privately owned it belonged to J. Place & Son Ltd, pipe manufacturers. The manager was called Henry Maden and the under manager was called Richard (Dick) King; this was the same Richard King that was the manager when it was nationalised and I worked there.

At the Moss Pit some of the colliers had decided to form a football team to play in the lowest division of the Darwen Amateur League, and after some practice matches I was chosen to play at right back. As the games were always played on Saturday, we had to jump on the tubs down to Bert's Brew, the pitch being at the Sett End not far from Bert's Brew. The trouble was, after work we were all knackered, and we never won a match all season – we once got beat nineteen goals to nil. However we once got our name in the *Last Sports*, a late Saturday night sports paper. It complemented us by saying 'that we were all good sports to keep turning up'.

One day when I got home there seemed to be a strange kind of atmosphere in the house. My mum was in bed, as she had been for a long time, and she had lost a lot of weight. Dad was there, which was unusual because he was generally at work, and Fanny was there. My dad said, 'Go to the pictures lad'. I told him I had no money, and he said he would pay. When I got home that night, something seemed strange about the living room, and then it dawned on me the bed had gone and so had Mum. Fanny was crying and the room was full of strange people. I asked Fanny where Mum was and who were these strangers? She told me that Mum had passed away, and that all these 'strangers' were her cousins and sisters. Mum was buried and life went on.

At this time there was a boxing competition, just amongst the pit lads. I entered one of the competitions with two mates, Frank Greenhalgh and Des Moody. The man who trained us, Fred Walkden, worked on the pit top, and there was something wrong with one of his eyes. It turned out that Fred used to be a decent boxer, but once his opponent's lace on his glove became undone and it whipped across Fred's eye, causing the damage. We trained in the loft of a stable at the back of the Bowling Green pub which was on Bolton Road, Darwen. This meant that I had to get home from the pit, get changed, then go to Darwen on my pushbike.

Our first fight was at Wigan. The only trouble being we had to work down the pit all day and then come up, go home and get changed and be back at Darwen centre for six o'clock to be picked up by a Coal Board van and taken to the venue. The lads from the other pits got the day off. When we arrived it was just one big changing room. They put your name down and who you were fighting on a blackboard. Des Moody was on first, he got beat easily. Then it was me, I was fighting a lad from a pit at St Helens, his pit was called the Clock Face Colliery. He beat me on points, Frank got a bye to the final, because only two had entered at his weight.

When I was eighteen I started my face training at the pit, I already knew a lot about it, through watching the other colliers. One Saturday morning we were working when the wheel end operator came up to the colliers and said that they had to go down to the wheel end and await further instructions. We were all waiting at the wheel end when the phone rang. The message was to get out of the pit as quickly as possible but don't panic. We got to the bottom of the drift only to find that there had been a roof fall in the tunnel, which provided the circuit for the air, so no air was getting down the pit.

A while after I had completed, and passed, my face training, there was an accident close to the bottom of the drift. Two colliers were working in some old working, salvaging that part of the pit. They took out what coal was left then propped the whole area up, sawed the props halfway through and the weight of the roof slowly broke the props, thereby taking the weight off the rest of the mine. The lad drawing for them was a sixteen-year-old named either Raymond, or Roland, Fox. When the colliers had filled a tub the young lad took it to the bottom of the drift. The men at the bottom of the drift would then send the tubs to the surface with the rest. There were only two men who worked at the bottom of the drift, one was named Walmsley Proctor Kay, nicknamed 'Wally', the other was the charge-hand named Colin. Young Fox then took an empty tub back, and because he could not see over the tub he had to keep his head down and behind the tub. All of a sudden he came to a grinding halt and when he looked round the tub all he could see was a wall of rock. A large piece of rock had come down out of the roof and it had crushed the two men to death. I believe that young lad Fox never went down the pit again. I know I never saw him again. I knew the two men who were killed, I had worked with both of them on different occasions, they were called Raymond Peet and Ernie Drinkall, and I was just twenty years old at this time.

I had been married for a year at this time, and my wife Helen was expecting. My wife had our baby, a little girl, and we named her Susan. I was working on the face now, so was Joe Walsh and Bernard Brooks, the two lads I had done part of my training with at Bank Hall. Rumours about Lord Robens closing the Moss Pit were becoming more and more common, because he had started ordering the closures of some other collieries. Most of the colliers believed that the Moss Pit was on his hit list. I had just turned twenty-two years of age when the order came that Moss Pit had to close. I was only young, and jobs outside the pit at that time were fairly plentiful. I felt sorry for the older colliers because the pit was the only life that they had known. I decided to get a job before the Moss Pit finally closed, and I got a job at Woods Brickyard in Blackburn. What became of most of the lads from the Moss Pit I never really knew. The NCB did gave you an option of going to another pit, but a lot of the lads didn't want this because they knew whichever pit they went to they would be on either regular nights or regular afternoons. Bill Fenton and a few more went to Huncoat Colliery. Alexander (Sandy) Mitchell, the under manager, went to Hapton Valley, as a training officer I think. Jimmy Brighty, the first collier I worked under, had died a few years earlier. Bob Eccles, the 'kind' fellow who washed my back on my first day at Moss Pit, became a landlord of a pub in Darwen.

If the Moss Pit had taught me one thing it was that I was never afraid of hard work. I found it hard adapting to life outside the Moss Pit. After finding work at other places I went working for the Gas Board, and then decided on a change so I went hod-carrying for a while. That was all right in the summer, but in winter not too good, and a lot of the time I was being laid off. So I went back on the Gas Board, this time as a ganger. I stayed there for another twenty-six years. They came round with a voluntary redundancy package, which in my case was too good to refuse. The job that I had at that time was coming to an end anyway, so I took the money. I moved to Knott End for nearly four years until the break-up of my second marriage. Helen, my first wife, had sadly died of a brain haemorrhage about fifteen years previously. I now have a dog, which was a rescue dog, from the RSPCA. I have had him for thirteen years – a wonderful and uncomplaining companion. I now have three daughters and seven grandchildren. 'Ben' the dog and myself now live in a residential caravan on the Shireburn Park at Clitheroe, and if I had my time over again I would still choose the pit for my job, and I would still choose the Hoddlesden Moss Colliery as my choice of pit – I have no regrets about that.

Huncoat Colliery

Huncoat Colliery was located to the north of the village of Huncoat between Accrington and Burnley. The two shafts here were sunk in 1893 to a depth of 276 yards by George Hargreaves Collieries Ltd. The Upper Mountain Mine was developed in the late 1930s, and during a reorganisation at the colliery in 1950 a new pit bottom was constructed at this seam at a depth of 219 yards. Two of the faces were fully mechanised using Anderton Shearer Loaders, which cut and loaded the coal automatically. Coal was transported in 3-ton mine cars hauled by diesel locomotives. Half the output went to general industry and half to a coal-crushing machine on the surface before being sent to the National Coal Board coke works at Altham. New pithead baths were opened in January 1953 by Colonel Bolton, Chairman of the National Coal Board's North West Division, at a cost of £30,000. The colliery was closed down on 9 February 1968 when economic workable reserves were exhausted, and the numerous faulting made further investment impractical.

Bill Lonsdale
Age at interview 83
Years in mining 22
Collieries worked at Huncoat, Accrington

I met Bill at his house in Great Harwood where I had arranged to meet him following a letter he sent to me regarding coal-mining memories. In spite of his eighty-three years, Bill has a marvellous memory of his coalmining days both locally and abroad, including a rather 'hairy' situation he found himself in while in Iran. Today, Bill enjoys his retirement meeting his friends and going out walking, and making little objects of furniture such as wooden reproduction rocking chairs, which he gives to his grandchildren.

Huncoat Colliery, Huncoat, near Accrington, seen here from the pit stock pile in the early 1950s. The pit was owned prior to nationalisation by George Hargreaves Collieries, being the first complete shaft sinking undertaken by this firm. Many attribute the first shaft sinking to the Scaitcliffe Colliery, but this firm only deepened the shafts there.

I WAS BORN in Accrington, almost in sight of the old Scaitcliffe Colliery, on 8 August 1923. My father was James, a council worker, and my mother was Margaret, *née* Gormaly. She was of Irish descent. During the wars, my mother was the midwife for Rishton and part of Clayton-le-Moors. I do not think my childhood days were particularly hard. We lived in Mason Street in Accrington. One day there was a heck of a commotion outside in the street. I think I would be aged nine or ten years old at that time. I asked my mother what was going on, and she said, 'That Jimmy Cropper started work at the pit today, and they left him down there'. Everyone was worrying about what had happened to him. I suppose they must have got him out though, they would not leave him down there. I was educated at St Anne's School in Accrington, and the last couple of years at the Rhydings School in Oswaldtwistle. My fourteenth birthday would have occurred during the school holidays, so I left school before then while I was still thirteen years of age. My first job was labouring in a brickyard called Clarke's Firebrick Works in Farnworth Road at Rishton, wheeling bricks around in a barrow when I was fourteen. I did this for a couple of years. The Firebricks Works was on the canal bank, and got their fireclay from Rishton Colliery, and of course when that closed down the Firebricks Works closed. Then of course the war started in 1939, and things changed completely then. I did some work at one or two little engineering works around town, and then got a job at the big

Bristol Aeroplane Company at Clayton-le-Moors. There must have been five or six thousand people working there. After the war it became the English Electrical Company. It was about this time I was acquainted with a lad called Alec Bray; he was an apprentice mechanic at the old Rishton Colliery at Rishton. After I had finished work I used to go to the pit and help out in the mechanics' shop or the blacksmiths' shop – anything that was going. I remember going down the Rishton Pit carrying a bucket of hot oil – I did not work there, I was just helping out. The bucket of hot oil had a bearing in it, and the hot oil kept it slightly expanded, not much but just enough to get it onto the axle bar. The mechanics on this job came from Mather & Platts at Manchester, pump engineers. One time, Alex Bray and myself took a girl down the Rishton Pit, after it had stopped producing coal. I would have been about seventeen or eighteen years old then. They still had a boiler in steam because there was still some traffic in the shaft, pulling things out. You know how things are when you are young: we asked these girls if they wanted to go down the pit – there were two of them, but only one came. The Bray family managed the Rishton Pit, four or five of them. Alec Bray's grandfather, who was managing the pit at that time, wound us down. We were actually going down to fix a gasket or something and we roped this young lass in with us. We took her under Whitebirk, quite a long way in, on a sledge – we were about an hour. Alec was bigger than I was, and he pushed her along on the sledge in front of him. She never said a great deal, she was quiet all the way in; it would quieten anyone I suppose!

When I was about twenty, or twenty-one, I joined the army for three years in the Royal Mechanical Engineers, as a vehicle mechanic. I went to France and Germany, the war was still on at this time, although it was in the closing stages. After the war I got employment at the old Moorfield Colliery, or 'Dicky Bridge' as they called it locally, working on the surface at the coke works, which was there then.

Because of my experience in the army and working at the various engineering works, I was put on maintaining the engines at the coke works – this and other work. The coke works was a very large place, the Germans knew about it and attempted to bomb it during the war. It was all lit up at night with the coke ovens burning. A lot of the coke went to Howard & Bulough's in Accrington, and of course gas was produced at the coke works. I remember that if we worked on the gas engines we were very ill indeed, from the gas – I was only at Moorfield Coke Works a couple of years. After this I went to the Huncoat Colliery at Huncoat; the manager here was Jack Whittaker, well he was the agent actually, all those pits around there came under his control. The under manager was a man called Billy Oldroyd. Jack

Bill Lonsdale relaxes after our talk about his coalmining days at his house in Great Harwood, Lancashire. (Jack Nadin)

Whittaker was a great man, he used to come in the mechanics shop – he was making a little engine or something – and he was always coming in to have his tools ground up. I was living in Clayton-le-Moors at this time, and I would walk to the pit over Whinney Hill Lane. I do not know whether I can remember the old Whinney Hill Pit, I think I can remember the engine house there, but I think there was a big compressor in it, I am not too sure. My wages at Huncoat were about £15, but we worked a lot of overtime, and I could get this up to about £18. This was quite a decent wage, it was a lot more than I got in the army, and I had a wife and two kids by then. Huncoat was one of the pits that were chosen by the new National Coal Board for investment, and work was begun in 1950 on reorganisation at the pit.

One of the biggest jobs was getting the Hunslet diesel locomotives down the shaft. I was there when they were getting the first one down, Vincent Whittam and me. We were at the shaft bottom, and we saw it at the top of the shaft. It was coming down like a yo-yo, coming down, and going back up as the rope took the strain. We had two chain blocks laid out on the track at the pit bottom; as the locos appeared at the bottom we were able to drag it in using the chain blocks. I do not think I was there when the second loco went down – but I was definitely there when the first one went down. Another major problem, although it was anticipated, was the fact that the larger-capacity cages would collide in the shafts without alteration. It was decided to move one of the headgear pulley wheels to one side, to give them room to pass, which was easier said than done. The pulley had been pressed on, and secured by four keys; it proved to be beyond our capabilities. The pulley had to be sent to Stotts of Haslingden by police escort and have the centre re-bored. Then of course the shafting was too small and a new one had to be made. We had a shafting to hand and a large lathe. Andy McGlugage, a former Burnley footballer, did the turning, and he used to play in goal I think. In the end it all turned out alright. Vincent Whittam and I were senior mechanics there, another was Tommy Finch, and they both helped me get used to mining ways and that sort of thing. Later I had my own apprentice named Bert Unthank; he later became an engineer of good standing at the Mullard's Works.

At this point I showed Bill some photographs of the Hunslet diesel locos underground at Huncoat featured in one of my previous books, Collieries of North East Lancashire.

Eh. Can you see that kink in the rails there? I laid those rails. I should have got a fireman to put me a line on, but I laid them anyway, that is why that kink is there.

It was quite amazing really that they managed to keep up the coal production while all this reorganisation was ongoing. One time while they were driving a little roadway at Huncoat they came across a big fossil tree, we got bits of it out, but as far as I know it is still there. There were many fossils at Huncoat, fossil ferns and things like that. A bloke called Tommy Smith once found a fossil duck, but this was at Deerplay Pit. It was very, very clear. Its head had been turned round, and it was flat. I do not know what has happened to Tommy now, he was older than we were, and he only went to Moorfield, he never came to Huncoat. Another man I remember at Huncoat was Jimmy Cregg, a fireman, he was very well-read, and he had studied the coal seams and geology. The first time I went underground at Huncoat, they wound very fast there. In the shafts it could be a little frightening. I was at Huncoat when we got word that there was some problem on the face or other at Scaitcliffe Pit in Accrington. When I got to Scaitcliffe I think the under manager was there, and I went down with him. We sledged in to the coalface, I followed him in, and it is five miles you know. The first mile you had to trap the rope onto your sledge, and that took you in. It was also low, you had to keep your head down, but I had done it before, I was much fitter than I am now. It took us about an hour to get to the coalface, kicking on the sleepers with our sledge. The tunnel went under the cemetery at Accrington, it followed the Burnley road. Coming out we got up some speed, it was all downhill, and they would all link up their sledges together. The problem at the face was a broken shaker conveyor, the bearing had gone, and there was nothing I could do about it. I would have to wait for the colliers to bring it out and get it to Huncoat to be repaired. My interest in machinery got the better of me here, and I was wandering round, looking at other machines. It was then I realised I was on my own – there was no one else about. So I thought where is my sledge? Although I had been down Scaitcliffe before, the under manager must have taken me somewhere I had never been. I could not find my sledge and my lamp was beginning to fail. Eventually I found the rails on the travelling road and began to crawl along until I came to a junction. The telephone rang here; they were ringing from the surface to find out where I had got to. I explained my predicament and they sent someone in to meet me as I was going out. My lamp was just a glimmer by this time – strange to say though, I was not frightened at any time, and I was down about six hours. By the time I had got out of the pit it was a cold winter's night and here I was, wet through and hungry. I was not going back to Huncoat, and I knew they would not let me on the bus in that state, so the only thing for it was to walk it back to Clayton-le-Moors, dirty, wet, hungry and tired. When I got home, there was no hot water, I could only heat a couple of pans of water on the stove and get myself washed down – I was back at work the day after though. I thought there must be something better in life, so I left Huncoat Colliery about 1960 or 1961, it was before they closed it down anyway. I had written to a mining firm in South Africa, and managed to get set on there in a copper mine in Rhodesia. I was married then of course, with two children, and the wife was not keen on going out there – so I went out there by myself for two years. Eventually they found us a house in South Africa with all the mod cons, hot water, bathroom, inside toilets, something we had never seen in Lancashire, and my wife came over to join me. From there I went on to work in various other mines in South Africa, including diamond mines and mines in Iran. While we were in Iran there was an uprising and the British Embassy advised all British citizens to leave the country. The mine we were at was a copper mine high up in the mountains, so high that we used to get nosebleeds. So we all had to get on board some buses; some of the men went one way, we went the other way.

We were travelling three or four days until we were put up in a hotel. We tried to contact the British Embassy here but it was closed. While we were going back to the hotel we were spotted by some Iranian rebels who chased us through the town with knives, but we managed to escape by dodging and turning through the traffic. Even when we got to the airport they were trying to overturn the buses – but we all got out of the country eventually. I'd had enough by then and retired when I was aged fifty-five, and now just enjoy life and my grandchildren, and take things easy.

∞

Norman Blackburn
Age at interview   69
Years in mining   c. 14
Collieries worked at   Scaitcliffe Colliery, Accrington, Fir Trees Colliery, Higham, near Burnley, and Huncoat Colliery, Huncoat near Accrington

Norman Blackburn kindly offered to contribute his mining memories following a plea from me in his local newspaper, the Accrington Observer. What follows is a transcript of the interview at his home in Huncoat near Accrington in late September 2006.

I WAS BORN on 25 April 1937 at Colne in East Lancashire, the son of Steven and Mary Ann Blackburn. My father Steven was a baker by trade, but he finished up labouring at the English Electric works till he retired. I was not at Colne long before we moved to Accrington while I was still a child – I was educated here at Accrington, at the Wood Nook Primary School and then the Accrington Technical College. Eric Barton was the head teacher at junior school. My favourite teacher there was Mrs Lang; she more or less looked after me while I was there. The year 1947 was a really bad winter. I was ten years old then, and Mrs Lang called at the house for me and took me into school. There was only me and the caretaker at school that day, and he only lived across the road – I never missed a day at school. I left school at the age of fifteen years and went straight into coalmining, doing my mandatory sixteen weeks' training at the Bank Hall Colliery at Burnley. The main man here was Ivor Jeremiah; he was the head of all the trainees at Bank Hall. One of the training officers I remember was Dennis Finn; he lived in Accrington near us. The wages for this training period was £3 3s, which was a lot when it was compared to what my mates were getting at this time. To get to Bank Hall I would catch the train and get

Norman Blackburn in his front garden at Huncoat on 2 October 2006. (Jack Nadin)

off at Burnley central station, and then walk over to the pit. The training consisted of half and half, one week at the pit and the other at a little school on Red Lion Street in Burnley centre. They built the old Burnley Building Society on the site when the school was knocked down. Most of the time while we were supposed to be training they sent us up to the park playing football, or running around the park. When they were making the playing fields where the Bank Hall Miners Social Club was, they sent us trainees to level it out. It was all built on spoil from the pit, an old slag heap from the pit. They sent us down the pit itself a few times at Bank Hall, on a training face – I remember having a go at those guns, those pneumatic guns down there. That was hard work.

After my training at Bank Hall Pit they sent me to Scaitcliffe Colliery which was almost in the centre of Accrington. The Scaitcliffe Colliery was only a small pit, and it was not very deep and was next to the old public swimming baths at Accrington. The signalling system at the shaft consisted of a wire going down the pit which connected with a hammer and a large iron plate. The banksman would pull the wire which would then ring at the pit bottom and in the engine house. You could hear these signals long before you even got to the pit – God knows what the people living nearby thought of it! I lived at the bottom of Hollins Lane at Accrington then. You could hear it all day – clang-clang – while they were winding coal. The coal went out of the pit yard on big motor wagons and was distributed all over; it was screened at the pit before it went out. John Tattersall worked in screens I remember. On one occasion one of the men I recall was short paid in his wage packet, and so he went to the office and said 'Not enough paper'. It was his way of saying there were not enough pound notes in his wage. They were paid in pound notes then of course, and his wage was made up. Another time, one of the workers brought his son to the pit, and went to see the under manger, Jack Acornley, about getting him set on at the pit. The father explained that the lad was a good worker and strong with it, but he was a bit backward mentally. 'That's all reight,' said the under manager. 'Looking round pit top, can thy see anybody else that's not the same round theer?'

There were no showers at Scaitcliffe when I first started, and I had to go home in my pit muck. I used to get washed in a dolly tub in front of the fire. I lived in a cottage at the bottom of Hollins Lane with one of those big cast-iron fireplaces with a boiler at the side. My mother used to ladle the hot water out into the dolly tub for me to get washed in front of the fire. There was a canteen though at Scaitcliffe; Mrs Riley was the canteen woman. At the bottom of the pit shaft it was all brick-arched tunnels, lit up by electric lights for about 30 yards, and from here into the pit itself was the chain-driven jig roads, and as the roads changed directions there were the places we knew as the 'gearings'. By the time the pit was finishing there were up to four or five of these gearings. The haulage was worked on the endless electric-driven chain system, the chain dropping into a fork on top of the tubs which was then propelled along, empties on one side of the track, full tubs on the other side. The height of the tubs determined the height of the roadways, and the tubs were about 3ft high. So the roadways too were 3ft high. The main jig road went parallel underground to Burnley Road in Accrington, and some of the working finished up at the back of Hambledon Hill. To get to the coalface involved going into the 'travelling roads' which were brick-arched for about 1,000 yards on an endless rope. This was up to the place they called 'The Long Gearing' and from here we had to go further into the pit on little trams. We used the trams here by trapping the endless rope onto our tram with a wooden flap attached to the tram which then pulled us in.

They started me off at first on the pit top for a few weeks in the screens. My first job at Scaitcliffe down the pit was at the shaft bottom, where I was put loading the tubs into the cages. It was always cold and wet here; it was always wet at the bottom of the shaft. I managed to get an old duffle coat off one of the mechanics to keep me warm. We had no rats or mice at Scaitcliffe until we broke into the Thornybank workings near Hapton. Then I was put on coal drawing for the colliers on the development, pushing tubs of coal in a tunnel 3ft high. Later I went as a timber lad, taking props, boards and things up to the coalface for the colliers to use. There was no haulage engine for this; once we had loaded up the bogie we had to push it physically along the rails up to the coalface. The bogie was a flat bogie, with steel bars up the sides, and we would put all the boards at the side of these to form something like a box, and everything else went in the middle, such as the props. We did improvise once though. I was working with Brian Maden, the fireman was Charlie Bond and we came across an old coal cutter which had been left there at the side of the tunnel. Charlie Bond and a mechanic managed to get it working for us, and we put a chain round the winding drum and used the cutter to pull the tubs up for us. Scaitcliffe came under the same management as the Huncoat and Calder pits, and the manager of this group of collieries was Bill Oldroyd, the under manager was a fellow called Jack Acornley. There was also an under manager named Stan Bollard later. One time we were talking about a football game. We asked Stan what he thought about it, and he said, 'I don't want you lot talking about football, get talking about coil'. Stan Bollard lived up Wheatley Lane way I seem to remember. The other firemen I remember here, at Scaitcliffe, were Norman Streets, Raymond Bannister and then after him, I was under Charlie Bond on the development. This was a longwall face development, using cutters to undercut the coal, which was then fired down using explosives. There was also Norman Williams; he was a fireman on the day shift. While I was on the development there was a fatal accident. A man was making the first cut with the coal cutter in the stable when all the rock above came down upon him and killed him. His name was Charlie Silverwood; he came from up Oswaldtwistle way. He should not have been on really, but one man was off work so they set him on as second man on the cutter. We had to stretcher him out of the pit, and it was a long way in. He was alive but very badly injured; he died as we got him into the cage. On a number of occasions we had to come out of the Scaitcliffe Pit through the Huncoat workings, but it was a long way underground. We had to go up through the 'Long Gearing' at Scaitcliffe and through an air door. This led into the old workings, all brick-arched, which led underground to Huncoat in the Rise District near a place they called the 'Odeon'. They called it the 'Odeon' because it took that long to build, it took longer than the Odeon at Accrington, it was massive. This was when the winding engine broke down at Scaitcliffe. The winder at Scaitcliffe was a steam winder. You could also go from Huncoat Pit to Calder Pit, and the Moorfield Pit at Altham, although I never went; these pits were shut down then. I was at Scaitcliffe from about 1952 until it shut in July 1962, ten years in all. By this time I had gone through the system from collier through to getting my shotfirer's certificate; this involved coursework at Burnley College which lasted about a month – I went with Brian Maden. They said that they wanted a shotfirer at Fir Trees Colliery, a little drift mine near Higham village on the outskirts of Burnley, so I went there.

Up to this time I had been working in seams 18in to about 24in high, and at Fir Trees the seam was about 4ft 8in high – it was like heaven to me! I saw the under manager Bob Kennedy, who set me on at the pit on what I thought were spilt shifts. But he said no, you can go down at ten o'clock at morning, do the job and come up again when you have finished. That will do me I thought. The manager at Fir Trees was still Bill Oldroyd, like at Huncoat and Scaitcliffe. Stan Bollard became the under manager at Fir Trees later on. It was good coal at Fir Trees, we worked the Arley Mine but it was a steep seam to start off with. The drift itself was also very steep, they tried to install a conveyor belt, but it was too steep, the coal just kept sliding back on itself. Eventually they installed a mine car and got the coal out up the drift with that. The mine car was hauled up the drift and straight on above a bunker – a door was opened at the back of the mine car and all the coal then slid out into the bunker. Coal was that good here it did not need any screening. The output was taken away in Coal Board wagons. Later they put a manrider in the drift for the men to go down. There were no showers at Fir Trees Pit at this time, although they did get some later on. I used to go to the pit on my motorbike and as I was living in Clayton-le-Moors at this time I would ride to the coke works at the old Moorfield Colliery and use the baths there. The method of working at Fir Trees was one of modified pillar and stall; the stall would be gunned out in the middle and then fired at either side. Where they had gunned out in the middle of the coal, this made the coal 'give' rather than having to fire it down with explosives in the solid coal. It was not unusual though for colliers to get me to fire down the coal where they were supposed to get it out with the guns. This was all unofficial of course. I was at Fir Trees Pit until that shut on 11 March 1966, and I was transferred to Huncoat Colliery after this.

It was like a large railway station underground at the shaft bottom at Huncoat, with the large loco engines, sidings and big tubs, quite a change for the little Fir Trees Pit. I was still on shotfiring and doing deputy work at Huncoat. Johnny Oldham was an overman here, Jimmy Cregg was the undermanager. Jimmy Cregg asked me to hang back one time when I was on the back shift, because they had some policemen coming on a visit. 'Show them round pit and be careful with them', said Jimmy. The policemen arrived in a Black Maria and all piled out. I got them all changed into pit gear, lamps, helmets and all, then I heard one of them say, 'Come on let's go and see how these so-called miners make all their money'. Right, I thought, I got them down the pit, and told them we were going up the Rise District, a good few miles of tunnel which were about 3 or 4ft high. Then I took them through the coalface, just to rub it in a bit, and then back down the tailgate. I didn't tell them that there was a manrider for all this, and when we all got out I can tell you I was knackered – but you should have seen them! 'That's how the so-called miners make all their money,' I said to them, 'and you lot only had to walk round, we have to do a full shift'. Some of the policemen could hardly stand up.

I was at Huncoat until that closed down on 9 February 1968. The men actually voted for the pit to close even though Joe Gormley, the Union man, tried to save the pit. This was a time of pit closures, and the men were just fed up of being moved around all the time – they were just sick of it, so they voted for it to close. They were just becoming industrial nomads shifting around all over the place. We could see anyway that there was no future in mining, they were closing pits left, right and centre.

I then went to the Florence Colliery at Stoke along with Norman Streets to have a look at the place, but there were no houses ready for us. So they put us in accommodation, but I did not like it. The pit was really deep, the faces there were about 12ft high, and it was hot, the men carried two-gallon water bottles to keep them going, it was that hot. They were working in just shorts, even underpants. The pit was regularly producing a million tons a year – all the pits in Burnley would not even produce a fraction of that sort of output. It was not for me, though, and when they came to say that a house was available, I said cancel it, and came back home. When it came to collecting my final wage, they even charged me for a shotfiring cable that I had left underground at the pit – I was disgusted. When I came back home, I was offered a job as ventilation officer at Hapton Valley, but I said no, I wanted to get out of the industry altogether. Government retraining schemes were in operation then, so I went on a course doing welding. I then went doing welding at various firms around until my retirement in 1997. I spend my time now just pottering about in the garden, and working as a steward at the Accrington Stanley Social Club. I enjoyed my time in the coalmining industry, and I still have mates I used to work with at the pit – good mates. I have known blokes since who I would not give the time of day to – pit lads were the best. I bumped into one the other day and we stood there nattering twenty minutes – the wife was pulling me along trying to get away. There was a comradeship in mining, you all looked after one another, you had to, that is the way it was.

∞

David Hargreaves
Age at interview   65
Years in mining   9
Collieries worked at   *Hoddlesden, near Darwen, Huncoat, Huncoat near Accrington*

David Hargreaves sent me his coalmining memories by email following a plea for information on the Internet.

I AM NOW sixty-five years old. I started at fifteen in 1955 as a trainee at Bank Hall. My first pit was Hoddlesden, a humble little wet drift mine on the moors above Darwen. I worked as a drawer in the upper Mountain Seam, 26in high, no ripping and two miles to the face on flat wagons. My sister's boyfriend was a night shift cutter at Huncoat; his name was Johnny Nelaney. He suggested I try there. So I went to see Bob Brooks, the under manager. That was January 1957. He started me on the afternoon shift on supplies and sometimes filling the mine cars on the loco level. The Upper Mountain Rise side was working (face nos 9, 11, 14), the flight loader district was being developed and the next one was to be the dip side going towards Calder Colliery. I was on day release every Tuesday at Burnley Tech, which was a nice break. I also joined the junior first aid team, which was about to become very successful. So they let me go back on day shift. The other team members included Joe Thompson, Geoff Ellis and Gerard Oughton. Gerard went on to have a long career in management in the South Lancashire coalfield.

We regularly won the NW Division finals in the 1958-62 periods. Our crowning glory was coming third in the National Finals at Blackpool, *c.*1962. There were eight finalists out of about 500 teams. The mine manager, Bill Oldroyd, was very proud of us. He treated us to a weekend in Blackpool. Still, my day job was as a timber drawer on No.9 District. My fellow drawers included Ken Scholes (known as Harry because his dad, Harry, was a collier), George Lewis and a big fat lad called Jimmy Butcher. Jimmy's parents kept a pub, the Veteran, in Blakey Moor, Blackburn. One day we were having our bait when Bob Brooks, the under manager, came upon us. He told Butcher he had been in their pub on Saturday night. The conversation went something like this:

Bob: 'Were that thi mother behind t' bar?'

Jimmy: 'Yesst, why?'

Bob: 'She IS an ugly bugger.'

Now Bob had once had a bad accident to his nose and it had been poorly repaired with a long straight split right down it.

Jimmy: 'Did you used to be a boxer, Bob?'

Bob: 'Aye, why?'

Jimmy: 'You must have been a bad 'un.'

We all laughed and Brooks went on his way up the board to the face. The tunnel undulated a lot and in the dips water accumulated. It was slippery and smelly. Brooks missed his footing and measured his length in the slime. Now we really laughed. Brooks regained his feet. He turned back to us and shouted, 'Butcher, nights on Monday!'

I recall our day wage pay as being about 30s per shift, with the bonus sixth shift for full attendance. As a boy you got an extra 5s per day every six months until you were eighteen and on full-day wage. Overtime was unlimited and you could work every Saturday at time-and-a-half. One week three of us worked a Sunday, which should have been double time, i.e. £3. When we went for our wages next Thursday we found we had only been paid 30s. I told the wages clerk and he spoke the dreaded words, 'You'll have to talk to Bob Brooks.' So the lads sent me in to bat. Bob threw me out of his office three days running, so, on the fourth, I pulled my master stroke. I went in there in my clean clothes, to show my determination. He rose to the bait, 'How much do I owe you?' 'Thirty shillings each, three of us.' He pondered for a moment then took half a crown out of his pocket. 'I'll toss you for it, double or quits.' Left with no choice, I called heads. It came down tails. He looked me full in the face and said, 'Right, now get your rags on and get down that effing hole!' It was at that point I realised my place in society. And I had to break the news to the other lads. But Bob could be generous, too. He gave a contract to me and the No.9 face timber man, Ken (I forget his last name), to run the TCR props and girders to the face every day. We did it in overtime (took about an hour) and got £3 a set, every day. It doubled my wages. Sadly, one day they moved Bob to Bank Hall and Jimmy Cregg took over. He dropped the price to 30s but I still did it.

During my coalface training, I almost lost my life. One afternoon shift we were put onto a job on the loco level, offloading girders at the rise side loop. There were two tracks running from the shaft to the far end and a set of points in bye feeding the loop. Before the points was a warning light and a STOP notice for the loco. The loop was full of empty 3-ton mine cars plus the heavy bogey carrying the girders. With me was my fellow trainee, David

Connor, and our trainer, Jimmy Taylor, an 'old man' of over sixty. A loco with a full load – probably twenty mine cars – came out from the far end, the Flight District (so called because they used flight-loaders). The loco driver was Brian Titterington and shunter Keith Starkie. They did not obey the stop signal. When they realised the points had been switched into the loop, they began shouting at us to run. They jumped out of the cab leaving the loco to smash into the empties, which began to charge past us as we were running towards the safety of the rise side loop. Jimmy Taylor was in front and was the slowest. I was in the middle. Then all the empties left the track and began coming at us in all directions.

David Connor was hit and came flying over my shoulder, taking my helmet with him. Jimmy made it to the loop, but slipped and fell. As I looked up, one mine car was coming straight at me. I remember thinking I was going to die and hoping it would not hurt much. Then it all stopped and went quiet except for David Connor screaming on the floor. It turns out he had broken his leg. Jimmy still had his lamp but mine had gone. We sat on a girder, he with his arm round me, saying he and I were brave, not screaming like Connor. I remember beginning to shake. It was a long time, or so it seemed, before we saw lights coming from the direction of the shaft. The under manager, Jimmy Cregg (he had replaced Bob Brooks who went to Bank Hall), and some others came and took us out of the pit. We went to Accrington Victoria Hospital where they examined us. I was physically OK, so I discharged myself and went home on the bus. Jimmy Taylor died a week or two later. I took the following day off and did not get paid for it. Today, I would have sued them for millions, but at eighteen you don't think about that. It was May 1959.

When I completed my training I went on the afternoon shift as a spare man. I did a variety of jobs including ripping and pan shifting on face One East and striking out on the Flight District. All the while I had been going to school on day release and when I passed my Ordinary National (1961) I was offered a place on the fast-track Sandwich Course at Wigan Tech for three years. I took it and passed my First Class Manager's Certificate in the November 1963 examination at age twenty-three. I think I must be one of the youngest ever to pass. You had to be twenty-three before you could sit and you had to have done seven years underground. I just made it. I went on to get many more qualifications, but that is the one of which I am most proud. To qualify you also had to have an O-Level English, a First Aid certificate, a gas testing and hearing certificate and to have passed the MQB examination in mining law. I left the mines in 1964 and emigrated to Canada, working as an engineer on an open-pit iron mine. Subsequent travels took me to South America and South Africa before coming back to the UK in 1974 to join a stockbroking firm in London. I now practice as a mining consultant. I shall be ever-grateful for my NCB training; it set me up for a great life.

∞

Joseph Thompson
Age at interview   66
Years in mining   12
Collieries worked at   *Huncoat Colliery, Huncoat, Hapton Valley Colliery, Burnley, Bank Hall Colliery, Burnley, Deerplay Colliery, Bacup, Scaitcliffe Colliery, Accrington*

I WAS BORN in Lancashire on 2 June 1940, the eldest son of Fred and Mary Thompson. I was actually born at the Rough Lea Maternity Hospital, located just outside Accrington, Lancashire. I am reliably informed that my first home

Huncoat Colliery Firefighting Team 1962–63. Left to right: Brian McCortney, Brian Maden, Geoff Ellis, Kevin Duxbury, Joe Thompson. (Joseph Thompson)

was a small, rented, terraced house in Rutland Street, off the Blackburn Road, Accrington, and was actually nearer to Church than Accrington. I have absolutely no recollections of living on Rutland Street during the first three or four years of my life, but do have a slight but hazy recollection of attending Spring Hill Primary School from the age of three. This school was at the top of the Spring Hill area of Accrington, just off Willows Lane, and in June 1943 the starting age was three.

I left Accrington Grammar School in June 1956, aged sixteen, with five O-Levels. I really didn't know what to do with the rest of my life, and to be honest I didn't really worry about it. I was healthy, very fit and active, a good runner, which I used as the training regime for my football, and could now look forward to earning my keep. I didn't have any strong job preferences but was certain that I did not want to do anything vaguely clerical. I didn't feel inclined towards anything 'safe', such as the civil service or working in a bank, even though it was always said that they provided the best pensions. I don't think that I even considered that I needed a pension until I was married. I certainly never gave it a moments thought at sixteen. I wanted to be seen to be bringing into the house some very welcome extra money and starting to pay for, and earn, my keep. I don't recall being fussy about what job I did but at the same time I must have ruled out trying for the classic apprenticeships in mechanical or electrical engineering.

There were plenty of engineering companies in the local Lancashire area and every year they recruited a good number of the local youths for their factories. Four- or five-year apprenticeships were nearly always offered to boys aged fifteen or sixteen. At this time, having a father, or a close relative, in an engineering company was as good a way as any of getting an offer of an apprenticeship. Eventually it was decided that I would try my luck with the National Coal Board. They ran all the nationalised pits in the country and in the Burnley area there were a good number of collieries, (known as 'pits'). At that time (1956) there were pits at Bank Hall in Burnley, Scaitcliffe Pit in Accrington, Thornybank Drift Mine at Hapton between Accrington and Burnley, Hapton Valley Drift Mine on the outskirts of Burnley, Huncoat Pit in Huncoat village and Deerplay Drift Mine on the moors above Rawtenstall and Bacup. In addition there were some smaller pits dotted about in the area, such as Reedley, Cliviger and others. My cousin Bernard already worked underground at Scaitcliffe Colliery. At my initial interview it was made clear to me that there were only a few ex-grammar school boys working underground in the pits. I quite fancied it, the money was good, and it became even more desirable when it was made clear to me that there were opportunities, in the future, for me to become a 'Student Mining Engineer', with all sorts of possibilities awaiting me should my studies be successful. They made it a clear objective for me that, should I 'pass' my medical examination and be successful in the basic underground training, I should start to study Mining Engineering on a 'day release' basis. 'Day release' meant me having one day per week at Burnley Municipal College, on full pay, studying mainly Mining Engineering Technology, Geology, Surveying, Mining Electrical and Mechanical Engineering and Extraction of Coal.

My mining career started at Bank Hall Colliery, which was located on the Nelson side of Burnley. It was a deep mine with underground access by cages suspended on steel winding ropes. I had passed my medical without problems, and on my first morning turned up at Bank Hall Colliery, which was the training centre for the Burnley area, to receive my new 'uniform'. It comprised a pit helmet, a 'boiler suit', pit boots with steel toecaps, a large belt, knee pads and most importantly a set of identification 'tags'. In addition to this equipment, you were issued with two lockers in the pit baths, one for your clean 'going home' clothes and one to store your dirty, underground clothes for the next day, plus two bath towels for use in the pit showers.

Identification tags, or checks as the miners called them, were extremely important to every miner. Mine were all numbered No.96. You had to hand one of these tags in at the lamp house in exchange for a cap lamp. When you were ready to enter the cage to go down the shaft, you handed a second tag to the 'banksman'. He then allowed you to enter the cage, closed and fastened the cage doors when it was full, and signalled electrically to the winding engineman, the man who actually controlled the winding engine. He would then start the winding engine and would lower you to the bottom of the shaft. It was at this stage that the importance of the identification tags became clear. The first tag was placed on the empty slot from where your numbered cap lamp had been removed. The lamp man could see that you had received a lamp and by the number proved who had been given that lamp. The second tag, which you gave to the 'banksman' before you went underground to do a shift's work, proved that you had gone underground. This double check was a fail-safe system of accounting for everyone underground, who they were, and how many were down the pit at any time.

I did not enjoy my first trip down the Bank Hall Mine shaft. It was over 1,000ft deep and cannot in any way be compared to a lift in a hotel. In a mineshaft the cage seems to drop totally freely, as if the steel rope had broken. You can see out of the cage and the wet slimy sides of the shaft simply become a blur as the speed increases. Believe me it really does drop quickly. You leave your stomach behind and swallow very hard to keep your

Joe Thompson, who now lives in Staffordshire.

breakfast where it should be. On my first descent I was with the other 'new starters' so any show of fear, or screaming out loud, was not an option. I just smiled, swallowed hard and tried to look unconcerned, and above all, tried to look tough. As the cage gets nearer to the pit bottom it suddenly and rapidly decreases in speed until it almost comes to a stop and once again your breakfast is just behind your teeth. I had by now met some new mates, Lol Moon and Gerard Oughton. Gerard lived on Elizabeth Street, which was in the Spring Hill district of Accrington, almost opposite the Spring Hill Working Men's Club. He was an ex-student from Oswaldtwistle Tech and like me had a few GCEs. Lol lived near Spring Hill School, just off Willows Lane in Accrington. We became very good mates for many years and the three of us added a few more mates over the next few years, most important amongst these being my cousin Bernard and Raymond Clarke. The common bond, amongst the five of us, being that all of us worked 'down t' pit' and this bond lasted a long time, even after we left the bowels of the earth.

If my memory serves me well, I believe that Basic Underground Training lasted about six weeks and to work underground you had to complete this training. On completion of this training, and to register this achievement, you were given a Certificate of Proficiency. We were lucky; we passed and were sent to the pit nearest to our homes, which had vacancies for 'Trainee' underground miners. I was sent to work at Huncoat Pit, in the village of Huncoat, just outside of Accrington. There was a bus route from Church Town to Accrington and then we went to Huncoat on the train. I remember catching the bus at five minutes past six in the morning, on Union Road, with Walt Whewell, one of our neighbours, then waiting in the waiting room at Accrington station for the train to Burnley. The first stop on the Accrington to Burnley line was at Huncoat station, then a two-minute walk to the pit baths.

Hapton Valley Colliery face workers' outing to Fleetwood in 1966. Joe Thompson is kneeling front middle. (Joseph Thompson)

I will never forget my first shift underground at Huncoat Pit. I was working the morning shift, (7 a.m. to 3 p.m.) and was allocated into the gentle care of my 'mentor', Mick Duffy. 'Mick will look after you. You'll be OK. Don't worry', they said. Mick was a supply lad, came from Blackburn, and was responsible for delivering pit props to the coalface. To do this you loaded a steel trolley with as many pit props as it would carry. You then bent over, put your head against your trolley and pushed it along the rails for anything up to a mile, using your legs for forward motion, until you reached the coalface. The pit props you delivered today were used tomorrow. The roadway we worked on was called the 'Top Bord', and it was, as I have said, about a mile long and we, Mick and I, were the only ones working on it. This roadway's other purpose was to carry the used ventilation air away from the coalface and return it to the pit top. Believe me when I say that I could smell everything the coalface miners were eating that day, and, much less pleasantly, what they had eaten yesterday and parted with today. Mick, despite being only two or three years older than I was, was an old hand at this supply business and he was definitely hardened to the strenuous physical work needed to fulfill this task. I was just an absolute novice and exhaustion seemed to set in very quickly. I was a good runner and played a lot of football so my legs were relatively strong, but pushing a bogie fully laden with pit props was a real effort on my part.

Halfway through this first shift Mick announced that it was 'Bait-Time' or lunchtime. Everybody carried their sandwiches in a tin box, usually an old OXO tin to protect the contents from the mice. We washed our sandwiches down with a drink of water carried in a metal screw-topped ex-Army bottle, which we carried in and out of the pit tied onto our belt. Five minutes after starting to eat his sandwiches, Mick was asleep. I was too scared to shut my eyes, never mind sleep. Finally, at about 2 p.m., we had finished our stint and all our pit props had been delivered. We were at the coalface and about two-and-a-half miles from the bottom of the pit shaft. I was knackered, yet proud that I had done my bit for the day. My legs had gone completely from the physically demanding work and all I wanted was a hot shower in the pit baths and then bed.

The safety rules in force underground were always taken very seriously by all the miners. My initial training emphasised the importance of preserving your personal safety, and the safety of those around, you at all times. Protective clothing was provided and it was expected that you wore it. Smoking underground was strictly forbidden, and when you handed the banksman your numbered tag, before entering the cage for the descent into the pit, he would search you for any 'contraband fags, matches or lighters'. My career as a trainee mining engineer was about to start. I had been granted a 'day release' to attend Burnley Municipal College. This meant that I was working underground for four shifts and having one shift off work to go to college. In addition, day release included at least one evening class every week. I'm sure you'll agree that I was working and studying quite hard. The college was just outside Burnley town centre, up Colne Road. I did a one-year initial Certificate in Mining Engineering, then a two-year Ordinary National Certificate in Mining Engineering (ONC 1960), followed by a Higher National Certificate in Mining Engineering (HNC 1962). In both these 'Finals', I was the top mining student at the college and was presented with two Hayden-Nilos medals for 'Meritorious Academic achievement in Mining Engineering at Burnley Municipal College'. I can clearly recall that when I was presented with the second Hayden-Nilos medal, Susan was with me and I'm sure that we've still got the newspaper extract from the *Burnley Express* for that evening. After five years of hard graft I felt very proud of myself and both medals are still on display in my office at home. Each time I completed one of these finals, I honestly believed that I had reached my limit. Completing and passing the HNC exams, however, made me realise that I could go on, do better, get some more qualifications and become a fully qualified 'Degree Standard' mining engineer. This awareness meant that now I was highly motivated to continue.

The progress I was making in my studies was being noticed, and after about twelve months of general underground work, I progressed onto work in the more technical aspects of the day-to-day mining operation. Before this 'promotion' however (promotion in my eyes), I had progressed from supplying pit props with Mick Duffy to being the main gate, coalface and conveyor transfer attendant. I was very involved in this job; there were lots of people knocking about and it kept me extremely busy and I never had time to get bored or fed-up. I started to work in the Surveying Section at Huncoat Colliery, as a linesman. Our task, two of us, Ken King and I, was to ensure that roadways were driven straight and correctly aligned in a pre-determined direction. By using two reference points, plumb weights, lengths of string and blocks of chalk, we could project a 'correct' line. These straight, directional lines were used by the rippers to drive these roads on the afternoon shift. From time to time we would work with the trained surveyors, who came from the Burnley area offices at Altham to verify our work. They checked underground alignment and directions using theodolites, and in the office they checked the progress we had recorded onto the Mine Plans. I not only found it very interesting work but it also meant that we had a small office containing the current Mine Plans, which we had to keep up-to-date for the colliery manager. Even better than this was the fact that we were almost our own bosses, we knew what we had to do and we could go underground, and come out of the pit, whenever we felt like it, as long as our work was done. My next change of emphasis was when I became a member of the Ventilation Team, led by Fred Holden. Fred was about two or three years older than I was and he was also studying for his mining engineering qualifications. Fred and I became good colleagues whilst working together at Huncoat Pit. The Ventilation Team was part of the Colliery Safety Organisation, led by John Cohen, and it was this group which led me and others into the First Aid Team at Huncoat Colliery. This group opened my eyes to lots of new mining interests. John Pickering was the training manager at Huncoat Pit and he was always looking for ways to increase the visibility of Huncoat in competition with the other pits in the Burnley area. Under John's direction we started first aid teams, football teams and firefighting teams, and competed against the other pits. The structure, which was supported by the national headquarters of the NCB, meant that if you won at the Colliery level, you could compete in the Divisional Finals, and if you won at the Divisional Level you were eligible to compete at the NCB National Finals. In First Aid the senior team consisted of Frank Curley, Fred Holden, Brian Maden and Raymond Parker, trained by Jimmy Allen. The junior team consisted mainly of Gerard Oughton, myself, Geoff Ellis and Bernard Clarkson, also trained by Jimmy Allen and all directed by John Cohen. We were supported from the top in our endeavours by the colliery manager Bill Oldroyd and his under manager Jimmy Cregg. First Aid was our most successful operation and for about three years or perhaps more we won the Area and the Divisional titles. These wins qualified either the juniors, or the seniors, sometimes both, to represent Huncoat Pit at the NCB National First Aid Finals. Every pit in the country was eligible to compete. I competed twice at the Winter Gardens in Blackpool and once at the Miners' Convalescent Home in Skegness. We stayed in four-star hotels on the promenade at Blackpool and in Skegness. Fred Holden, Brian Maden and myself became Individual National Champions in Blackpool. These overall team results were a great honour for Huncoat Pit and our efforts were recognised at a celebration dinner thrown for the First Aid Teams by the Burnley area top brass. I still have the photograph of this event and I know it was important because, as well as all the colliery management and ourselves being in attendance, there were two senior Union officials sat at the top table. To get to this level we spent a lot of time training on the practical aspects of first aid at the local St John's Ambulance Drill Hall in Accrington, down Bull Bridge. We were almost word-perfect on the contents of the *St John's Ambulance First Aid Book* which enabled us to get high marks on the theoretical questions. We were obsessive and dedicated to winning because the prizes were good and we got plenty of time off work to practice. It has to be said that we also put a lot of our own time into becoming as proficient as we were, including competing far and wide as a team, even daring to venture over the border into Yorkshire. Even this sacrifice was worth it, if it helped to improve our performances, and it did. To give you a small idea, let me outline one of the first aid situations you could find yourself up against in a competition. I will try to describe my individual test one Saturday afternoon at the Winter Gardens in Blackpool. The curtains opened in front of an audience of about 250 experienced first-aiders from all over the country. I am off stage and cannot see what the audience can see. The compere informs everybody over the public address system that, as they can see, this light aircraft has just crash-landed and the pilot is hanging out of his cockpit; he is bleeding and unconscious. I've got five minutes to treat him and hopefully save his life, and whilst I'm doing this two or three adjudicators are closely watching everything I do and marking me accordingly. As I check his vital functions, I notice that there appears to be an arm, in a sleeve, on the ground. As I move him, it is obvious that he has only a torn and bleeding stump left in his jacket and he could bleed to death. Getting him eventually onto the ground, I move his leg and his boot becomes dislodged, leaving only the bottom part of his leg showing as a stump and its covered in blood. Closer inspection tells me that his foot is still in the boot and some distance away from the rest of him. To summarise before you faint, my patient has lost an arm and a leg and is bleeding to death. All such 'casualties' come from a very professional organisation called the 'Casualties Union' and my patient really had lost an arm and one of his legs before he came to Blackpool. By the application of a lot of make-up and copious lashings of tomato ketchup, his injuries looked extremely real and the final nice touch was the arm in a sleeve and his foot still in his boot.

Firefighting is another of the skills required by miners and we had, for a couple of years, a fairly strong team. Some of the names I can recall include Ken King, Kevin Duxbury, Brian McCartney, Cyril Stubbins, Brian Maden, Fred Holden, David Winterbottom, Gerard Oughton and myself. In 1957 and 1958 we were the Burnley Area champions and represented the area at the Divisional Finals. Firefighting competitions were all about putting out fires using a variety of fire tenders, hoses, fire extinguishers and other appliances, in both simulated underground situations and real pit-top scenarios, against the clock. Split seconds were vital and it was always a race against the clock. All the team members had to be fit, quick and mentally alert. For most of the time we had most of these qualities.

Huncoat Colliery football team competed in the NCB Knockout competitions during 1959 and 1960. Some of the names I recall include Geoff Ellis, Brian McCartney, George Floyd, Brian Maden, Norman Waddicor, Jeff Beech, Kevin Duxbury, Alan Parker and one or two whose names I've forgotten. I think the only claim to fame from amongst this lot was that Kevin Duxbury's son Mick played professional football for Manchester United and a few other top football league teams. All of us played for other teams as well and I know that Norman Waddicor played semi-pro for Darwen. We got the chance to play on some very good grounds. The bigger pits had first-class pitches and excellent facilities.

First aid, firefighting and football, on behalf of Huncoat Pit, were important interests during my late teens, but I still had my new job in the Ventilation Team to fulfil. It was a job that I found to be consuming and very interesting. We carried out all sorts of tests and checks on every individual aspect of mine ventilation, including anemometer readings of air velocity, air temperatures, air velocity gradients and variations, the humidity content of the mine air, the percentage of gases in the air, air speed and many other technical requirements. One specific test was to collect air samples for analysis back at the laboratory. We also worked with the supply, location and replenishment of the 'Stone Dust Barriers'. I remember Fred Holden putting three or four of these bags of stone dust on Frank Westwell's chest as a 'dare'. Frank, who was small and thin, went bright red then pale blue in the face before some kind soul lifted the bags off him and air returned to his lungs. You could say that this was another ventilation test.

One particular part of the job that I really liked was that we, at Huncoat, were responsible for all the air testing done at Scaitcliffe Colliery. Scaitcliffe Colliery was in Accrington, not far from the town centre and just above the swimming baths. On the required day, in the late 1950s, we would be taken in a van from Huncoat to Scaitcliffe with all our gear and go down the pit and carry out our work. One eventful day sticks in my mind. It was a Saturday and Brian McCartney and I were at Scaitcliffe, on overtime, carrying out the usual set of air readings. The face workings at Scaitcliffe were more than four miles from the shaft, and to get from the shaft to the face, and vice versa, meant a difficult journey. You had to kneel on a wooden sledge, with its four wheels running on steel rails, and push the sledge, and yourself, forwards along the tracks, using your legs, for the whole four-mile journey, which took about an hour. On our return journey, as we careered, illegally, in tandem, down a long slope, totally without brakes, Brian somehow became separated from his sledge and crashed head first into the wall. The roads that we were travelling along were about 4ft high and about 5ft wide, so there's not much room for error. I piled into Brian, who created a superb braking and cushioning effect. A close personal inspection showed that only Brian was injured, I was OK. The injury appeared to be some bones broken in his hand, nothing really serious. On our eventual arrival at the pit bottom, we were informed that we would have to ride in the standby shaft which was normally reserved for emergencies only. Brian and I were loaded into a cage not much bigger than a large bucket, which only came up to our waists. We were given a wooden bar with a loop on each end. The loops went round the guide ropes and we were told to hold this bar in position, to prevent the cage from swinging throughout our trip back to the pit top. We were in a bucket on the end of a rope!

It was around this time, when I was about eighteen or nineteen years of age, that I was playing the most football of my life. I can remember playing at King George's Playing Fields in Baxenden, White Ash Playing Fields in Oswaldtwistle and pitches in Clayton-le-Moors, Rishton, Accrington, Stanhill, Great Harwood, Blackburn and many others. During one particular season, I was playing for Accrington Collieries in the local First Division and also played for a Blackburn Rovers Junior Team in the local Senior Cup competition. The Accrington Collieries team was made up of players from Huncoat and Scaitcliffe pits and was playing to a good standard. I can remember playing in the morning in a snow storm and cutting both my knees, in a 'friendly' for a third team, and then playing seriously in the afternoon for the Collieries team in the local First Division. I loved to play and two matches in a day was no problem. The Collieries team used to meet in a local pub and as some of the players trained on 'Best Bitter', they would have a couple of pints to warm up before playing. This form of training must have done them good because they never got tired and we won more than we lost.

Before any miner is permitted to work on the coalface, it is a legal requirement that he is fully 'face trained'. To qualify for coalface training, you had to be at least eighteen years of age, and be considered to have satisfactorily completed a period of general underground work. I had completed two years, carried out a variety of tasks, and was now ready to progress on to working in some capacity on the coalface. I was eighteen years of age at this time. Huncoat Colliery had a coalface reserved specially for the training of eighteen- and nineteen-year-old miners. Everything related to this coalface was under the direction of the Colliery Training Manager. The face itself was run by a team of instructors who taught every aspect of coalface work, including drilling, ripping, coal cutting, panning, roof supports, roadway supports, packing, conveyor extending and, of course, shovelling the coal onto the conveyors. I seem to recall that this coalface training took about six months to complete and at the end of this training, if successful, you qualified to work on the face. In addition to this, I was chosen to be trained as a 'coal-cutter operator' by Walt Whewell, who was a neighbour of mine. The coal cutter cut a slot, 6ft in depth, into the coal at one end of the coalface and then dragged itself on a steel cable to the other end of the face, 150 yards away. As it progressed along the face, which was about 3ft high, it created a continuous slot, 2m deep into the coal. The coal cutter used a steel rotating chain filled with tungsten-tipped picks to rip out the coal and load it onto the face conveyor. A good few years later, this principle evolved into the Anderton Shearer Loaders of the 1960s.

At this time at Huncoat Colliery, colliers on the production faces and certainly we on the training face used wooden pit props and wooden roof supports. I became quite proficient at lying on my side, in a few inches of water, swinging a pickaxe-handled steel sledgehammer to knock a wooden pit prop tightly into a vertical position under a wooden roof support. At times I've seen pit props 'weep' as the roof weight compressed the prop and squeezed the moisture out of the wood. I've seen wooden pit props crack, with a sound like a gunshot, and bend over under the weight and I've seen them being pushed into solid rock floor when the weight becomes excessive. You only catch a fleeting glimpse of this happening, because a split second later, I was usually running on my knees as fast as I could for the safety of the nearest roadway. I was quick but I never got there first.

Being face trained and also cutter trained, I could now start to work as a fully trained miner. My reward was to be given a 'vacancy' on the night shift. I should really have known because the sequence of extracting coal, at that time, meant that the coal was cut on night shift, shotfiring took place next and the coal was removed onto the conveyors by the colliers on the morning shift. The afternoon shift moved everything forwards and 'ripped out' the roadways. The night shift then cut the coal again. I had been hoping for a vacancy for a face worker on maybe the afternoon shift. Thinking back though, maybe the management got it right; some of the jobs on afternoons were very demanding with extremely heavy lifting being required. Drilling holes in coal or rock was very tiring and I found it difficult enough to have to lift a drill with a 7ft drill on it, never mind push it into the rock. Some of these miners were built like the proverbial 'brick shithouse' and not all of them were 'gentle giants'. Overall I have nothing but the greatest admiration for miners and the work they did. I thought this at the time and I still do. My first few months were spent on the night shift as the 'spare' cutter man. When someone was absent or not available, I was the first reserve. Because of this, I would arrive just behind the rest of the team, and it was often said to me that they knew who they were getting as 'sub' without seeing my face, because they could hear me whistling as I walked towards them in the darkness. I must have been happy at my work; I know the money was good. The period of time I spent as a 'sub' was most beneficial to me and I learned a lot from some very experienced coal-cutter teams. I have some very good memories of finishing work at 6 a.m. on a Saturday morning, going to my Aunty Annie's house on my motorbike for breakfast with [my] cousin John Shorrock, who was also working on night shift, but at Holland's Pie Factory in Baxenden. Guess what we had for 'breakfast'?

Eventually I graduated into my own team: Mick Brown from Great Harwood was the cutter driver, Geoff Ellis from Oswaldtwistle was on the steel cable and I was on the electric power cable. This cable was as thick as a man's wrist and was about 100m in length. My job was to make sure that Mick had enough cable, and no more, at any time during the shift. I had also to ensure that we didn't damage this cable and the spare one we used at the end of our shift. Two other names which spring to mind from the night shift are 'Liverpool Jack' and 'Johnny Dog'; I never knew their real names. Coal cutting, as the name suggests, created a lot of dust as the cutter picks ripped out the coal. We were working in this dust all through the shift and I have vivid memories of being on the coalface, wearing a face mask filter and plastic goggles and watching the dust build up on my arms until it was almost an inch thick and then blowing it off. It was just like working in a black sandstorm. You could tell who were miners if you saw one of them outdoors, they always looked as if they were wearing black eye make-up. Throughout this period of working on nights, I continued to study at Burnley Municipal College. I had established that, for me, the best system was to have the night off before I went to college. This meant that I was fresh for my studies after a night in bed. I then went to college from 9 a.m. until 8 p.m., and then went to work at 10 p.m. for a night shift underground. I then slept about twelve hours the next day to recover. The alternative of working all night and then going to college was impossible; I was falling asleep in the classroom. The deal was one day at college in exchange for one night shift. The problem was that changing from nights to days and then back again, within three days, seriously disrupted my sleep patterns. By 1963 I had completed my HNC in Mining Engineering, had gained my Shotfirer's Certificate and by 1964 I had qualified for a colliery under manager's certificate, under legislation relating to the Mines and Quarries Act, 1954. The official title of the under manager's certificate was the Second Class Certificate of Competency for Mines of Coal, Stratified Ironstone, Shale or Fireclay. This certificate was issued by the

Minister of Power on the recommendations of the Mining Qualifications Board and it still hangs on my office wall, at home, to this day. I had by this time progressed from coal cutting to shotfiring and was working from 3 a.m. until 10.30 a.m. (Bed at 9 p.m., up at 2 a.m., then work, back to bed at 12 noon and out of bed at 3 p.m. This really does mix up your sleep patterns.) At this time (1963) I had exhausted the facilities of the Mining Department at Burnley. To progress further, I had to enrol at Wigan Mining College. I enrolled there with Peter Ashton who lived in Burnley and worked at Bank Hall Pit. Peter had a car and he was my transport from home to Wigan once every week for another four years. At Wigan Mining College I completed my academic and technical qualifications in mining by gaining a colliery manager's certificate, the First Class Certificate of Competency, and I became an Associate Member of the Institution of Mining Engineers (1967) and finally, in 1974, achieved my pinnacle and became a Chartered Engineer. I was a very proud young man. My father-in-law, however, kept my feet firmly on the ground. He said, and I quote, 'You'll be one of them clever buggers with lots of letters after your name'. He was right, I had studied part-time for ten consecutive years and I felt like I'd earned them. Was it worth it? Yes.

The next major change in my career came in 1965 when I transferred to Hapton Valley Colliery. I was a qualified colliery under manager, studying for my manager's 'ticket' and I felt that it was time to start my next 'career' as a manager. Hapton Valley had suffered an explosion a few years before I went there, so, unfortunately, I knew a bit about the place. The manager's name was Adam Weir. Other names I recall are Jimmy McKillop, who lived in Burnley, and Jimmy Holden, who lived in Oswaldtwistle. I was given the manager's job on the afternoon shift, with responsibility for that shift's foremen and their shotfirers. My title was 'shift overman' and the foremen/supervisors were titled 'firemen'. The three overmen, one on mornings, one on nights and me on afternoons, reported to the colliery under manager and he reported to Adam Weir, the colliery manager. I worked five afternoon shifts, Monday to Friday, and a Sunday morning, as and when required. Susan and I were married at this time and I well remember leaving home to catch the 12 noon bus from Accrington to Burnley, which passed the pit gates on the Burnley to Accrington road. At the end of my shift, I used to ring for a Coal Board van if I couldn't catch the last bus at 11 p.m. out of Burnley. The system of organising the work to be done at 'The Valley' was that every team had a 'team leader' who directed his own team in exactly what needed to be done. He would allocate work to men and would move some of them if there were more important jobs to be done. For this he received a weekly bonus, which was saved and put into a separate account to pay for a day's outing for his team. Every underground miner was in a team and was therefore included on these trips. In my time there, I was invited to go on the trip, along with the two other overmen and all the firemen and shotfirers.

The day started as Jimmy Holden and I got off the Burnley bus at five minutes to seven on a Saturday morning, outside one of the pubs on the outskirts of Burnley. It was a cold, wet morning and the pub was in darkness. Jimmy had obviously been before, because he knocked on the back door and in we went. Inside it was already in full swing, there were tables full of beer, tables full of food, crates of bottled beer on the floor ready for the coaches and drinking was in full flow. This was when I learned that Jimmy Holden didn't like beer; his preferred 'tipple' was a half pint of port. The second thing I learned about Jimmy, during that day, was that he liked a lot of it. I think that there were three coaches, we loaded the crates of beer and at about half past eight left the pub. We had been in the pub for around one and a half hours and we'd had plenty to drink and free food. Could it get any better? Yes it could. Everybody on each coach was given an envelope with 'expenses', including all the managers and supervisors. Our next stop was a large transport café on the road to the coast for a full English breakfast at 10 a.m. This was on top of the bottles of beer on the coach. The pubs in Fleetwood opened at 12 noon, so our coaches left the breakfast halt in time to catch opening time in Fleetwood. At 1 p.m. it was time for another free meal in the pub restaurant, followed by a quiet hour in the bar with a few more drinks. Three o'clock was the time for one half of the group to play the other half of the group at football on a great big pitch marked out on the sands. As I've said before, I was very proficient at avoiding leg-breaking tackles at football and in this match the opportunity for a bit of 'revenge' was obvious as was the chance to 'do' the managers. I survived. Exhaustion quickly set in amongst the majority, after all that food and drink, and the sands were soon littered with unconscious miners, sleeping off their boozy lunch. Some, however, kept the match going; they clearly had hollow legs. Before we knew what was happening, it was time to go for our tea. The match had obviously sharpened the appetites and our 'tea' was a full three-course evening meal, once again in the restaurant of one of the big pubs in Fleetwood. The coaches were outside of this pub and after a few drinks after the meal, we all piled back onto them for the journey home. The day was superbly organised and how it was paid for, and how much it cost, I have no idea. As a 'team-building exercise' it was brilliant, unfortunately team-building hadn't yet been invented!

I had my worst accident at Hapton Valley. I was buried under a roof fall – nothing too serious, just very scary and a damaged elbow.

I worked at Hapton Valley Colliery for two years between 1965 and 1967, working on the afternoon shift. In 1967 I was selected for what was called at the time the 'DPT Scheme'. DPT stood for 'Directed Practical Training' and it was really the National Coal Board's Senior Management Training Scheme. The programme lasted for a year and was designed to acquaint newly qualified mining engineers with the management of the collieries run by the NCB. Entry onto this course was very competitive, involved various interviews and selection procedures and required a number of personal references from colliery managers who could report on your suitability. Not every applicant was successful, you could be rejected. The course comprised a mixture of practical studies, colliery visits, weekend and week-long courses and a fixed number of residential courses at the NCB colleges at Chalfont St Giles in Buckinghamshire, Long Benton outside Newcastle, and in Edinburgh, Scotland. I had qualified as a member of the Manchester Geological and Mining Society and I found this to be advantageous whenever I was at Chalfont. Some of the permanent staff visited our meetings in Worsley at the Mines Rescue Station, just outside Manchester, and seeing a familiar face sometimes helps.

I spent a fair amount of time in Edinburgh, where the Scottish division of the NCB had a mining college. We visited Pontefract and Newcastle, visiting the Yorkshire pits and some of the pits under the North Sea at Easington in County Durham. I went down the salt mine at Winsford in Cheshire and I went down every pit in Lancashire. I submitted endless reports, went to innumerable meetings, I even presented a paper entitled 'The Installation of Powered Supports for Longwall Mining'. I met many young, and some not-so-young, mining engineers from all over the UK. I met Lord Robens, who was the chairman of the NCB, and remember being impressed when he had to restart a televised speech he was making to us. He simply started again and was word-perfect, very impressive. At the end of this interesting, yet difficult, year, I had made a lot of contacts and had learned a lot. It was a scholarship well worth winning.

The year was designed to prepare us for management positions within local collieries, and at the end of my year I was sent to Bank Hall as the 'deputy manager'. I was twenty-eight years old. Little did I know, at that time, that my career as a mining engineer would shortly grind to a halt. My time at Bank Hall Colliery was to be limited to ten months. I had, during this short period of time, a really interesting encounter with mining 'contractors', a breed of men I had never previously encountered. We had employed their parent company to drive some roadways through some difficult geological structures and at a fairly steep inclination. These men were almost self-employed and were paid, according to our contract with their company, 'on results'. If the road didn't progress, we didn't pay them. Our first surprise was when they arrived, prior to starting work, towing caravans behind their cars. They informed us that they would be living in these caravans and would be working seven days a week and twenty-four hours per day. The caravan beds never got cold, as one got out of bed, another one got in! Our second surprise was when we started to try to record their names, addresses and personal details. None of them was forthcoming! We were told to mind our own business, they claimed to be employees of their parent company, and none of them would give us any personal information. Without personal details, we would be unable to pay them as individuals. They had an answer to this little problem as well. We were to pay their charge hand the total amount due for the week, and he would split it up amongst them, which we did after much discussion with the contractor and our legal advisors.

My final surprise was one morning when I was sent to 'shotfire' the first tunnel entrance. When I arrived at the tunnel face, the contractors had drilled all the necessary 3m holes and had no time to lose. Their charge hand immediately took all my detonators off me and as his men helped him, he loaded the drilled holes full of detonators and dynamite and completed a complicated wiring circuit. I should clearly explain here that this was totally without my input, advice or assistance of any kind. It was also illegal and I would be unemployed if I was found out. As my career 'teetered on a knife edge' I was allowed to turn the key and the roadway exploded. The results were absolutely perfect! These men were experts at what they did, they knew exactly what to do and they taught me as well. They knew that if I made a mess of the 'firing' they were the ones who would suffer, not me. They were professionals and a real league of nations. Amongst them were Russians, Irish, Italians, Poles, English and a few 'unknowns'. I learned later that all of them had a common goal, not to be known to the tax man and, no matter what, they had no intention of becoming traceable, hence their reluctance to divulge personal details. We had absolutely no problems with any of them, they did a superb job.

One other experience springs to mind from my time at Bank Hall. One Sunday lunch we came up the pit and at that time we, the managers, had our own private shower block behind our offices. We had been working in a flooded area and all of us, including the colliery manager, were soaking wet. The manager, unlike the rest of us, decided that he would not shower, but would remove his wet clothes and immediately drive home in a dry pair of underpants. He left to do this and we immediately rang the police, to report having seen a naked man driving a car down Colne Road. It took him quite a while to explain to the local bobby why he was only wearing underpants, whilst driving home, on a Sunday.

I had decided to leave the mining industry. There did not appear to be a positive future for me and my family in this industry. I was working at Bank Hall Colliery, and was operating as a deputy colliery manager, on all sorts of different tasks from People to Production. One job I will always remember involved some illegal removal of coal from a drift mine just above Bacup on the moors. One of the Burnley coal seams comes to the surface on the sides of the valley. The road through the valley includes Rawtenstall, Haslingden, Stacksteads and Bacup. This seam had been extracted legally at Deerplay Drift Mine but this illegal pit was not in the scheme of things and this illegal mining was theft using a farmer's field high up on the sides of the valley. I was directed by the Burnley Area Production Director to go and close it down. I put a box of dynamite and a bag of detonators in the back of the pit van and set off for Bacup on my own. I did wonder about the Health and Safety legislation, but as it was an illegal operation, I felt it would be OK. I called at the local police station as agreed only to find it empty and locked. I simply carried on, found the entrance to the drift and set enough explosives to totally destroy the entrance and the workings, which had developed to about 10 yards into the seam and were about 3ft high. I shouted, 'Fire in the hold!' for the benefit of the local cows and up she went. There was an enormous bang and half the field shot into the sky in a big cloud of smoke. Job done, private mine destroyed, time to get back to the office. It was then that I heard the sounds of a police siren, then another, this time with blue flashing lights, followed by the distinctive sounds of two local fire engines. Many hours later, after a lot of explanations and telephone calls, I was allowed to leave and get back to my other duties.

Apart from these interesting interludes, my career was stagnating. It was the middle of 1968 and there were more qualified mine managers than there were mines for them to work in. In addition there was the constant threat of more colliery closures and even more surplus managers on the market. I could see that I was at the end of a queue and that queue was getting longer as the number of operational pits declined. In my opinion, the writing was on the wall, I could see it and I wanted out. Outside of Great Britain, my options in mining were quite varied. I was offered positions in Zambia, in South Africa, at Nchanga Consolidated Copper Mines, working at a depth of 2,521ft, with a total workforce of 9,700 men. But this didn't give us any assurances about our own long-term futures, so we turned it down. There were always plenty of mining jobs in Australia, from coalmining to opal mining, with gold mining in between. The problem with these jobs was always the same;

the mines were always hundreds of miles from civilization. Some provided 'barracks' type accommodation, where the daytime temperatures were often 125 degrees and above. Too hot to sunbathe and there wouldn't be much social life, living in the desert. In addition, the educational needs of our children could only be met by them attending boarding schools in the cities, hundreds of miles away from us. This was not acceptable to us and Australia is a long way from Accrington. We were ready to contemplate a complete change of location, but moving to the other side of the world seemed to be a step too far, so I registered with the Professional and Executive Register in Burnley and through them I was successful in obtaining a managerial position with an international company in Ireland. My family and I left the mining community and England behind and headed for what we felt were 'pastures green' (and they were).

Ian Carter
Age at interview   59
Years in mining   8
Collieries worked at   Bank Hall Colliery, Burnley, Huncoat Colliery, Huncoat near Accrington

Ian contacted me after an appeal in the local newspaper for mining memories – he told me that he was in the same group doing their training at Bank Hall as myself. I had to admit that Ian must have had a better memory than me about these matters – for although I have some recollections about training at Bank Hall Pit, they are very few. Nevertheless Ian was able to give some interesting reminiscences about his training at Bank Hall and his employment at the Huncoat Colliery following this. When Huncoat finished, Ian was transferred back to where he first began his mining career at Bank Hall Colliery and finished there two years before that pit closed down.

I WAS BORN in Blackburn in August 1947, the son of William and Ann Carter, née Dowty. Childhood days were pretty meagre in those days, with clogs and hand-me-down clothes. Dad would get paid on the Thursday, and Mum would straighten up at the corner shop and the next day start 'strapping' again to make ends meet until the following payday – but looking back I think everyone was in the same boat in those days. I went to school at St Mary's in Clayton-le-Moors and in 1959 I went to the Holy Family School at Accrington. I left school aged sixteen. My mother went to see Ivor Jeremiah the main training officer at Bank Hall Pit in Burnley; he lived

Ian Carter with some of his mining memorabilia at his home in Clayton-le-Moors in mid-November 2006. (Jack Nadin)

just around the corner from us. My mother was a friend of Ivor's wife. Mr Jeremiah was a man of high esteem locally, he would walk around town in his waistcoat greeting folk in his broad Welsh accent, and he was able to arrange for me to get set on as a trainee miner. To get to Bank Hall Pit, I had to catch the five-thirty bus every morning to get to Burnley in time for starting work. Burnley at this time always appeared to be a town covered in smog – a thick greenish fog that lingered everywhere. The early morning buses had greasy windows and to add to the drear almost everyone smoked at that time causing wet brown nicotine stains to run down the glass of the buses. At the pit we were put under some training officers, these included Herbert Higson and Jack Connell. Our training consisted of going to various other pits around the area, such as the recently closed Hoddlesden Colliery, Hapton Valley, Thorneybank and Old Meadows Colliery at Bacup. My first week's wage was £4 19s 7d. On other occasions we went down Bank Hall Pit via the No.1 shaft to work on a mock coalface, which had been set up especially for the trainees. At the bottom of the shaft we would turn left through some air doors. One thing that struck me here was a bright orange stream, which ran at the side of the roadway. If you put your hands in it you had brown or orange hands for the rest of the day.

The trainees were put to work doing a bit of ripping or setting arches and lashing chains on a moving haulage rope. Making manholes at the sides of the roadways was another task we had to do – ripping out the rock and setting small arches. Properly these were called refuge holes, a place where the men could go if, say, there was a runaway tub. After my training I was sent to Huncoat Colliery – this was just before Christmas, beginning work properly after the holidays. My first job at Huncoat was in the Flight District as a tackle lad, or supply man, taking equipment up to the coalface for the colliers and rippers. On the main manriding shaft at Huncoat you went around to the main travelling road, which was double tracked and all lit by fluorescent tubes, past the Dip Workings about halfway up, then on to pass the Rise Side Workings. A further 500–600 yards brought us to the place where they were taking all the coal out in retreat and there was a face on either side of this roadway. I remember the first day at Huncoat, when I was late in getting there, so instead of going down at seven o'clock I had to go down at eight o'clock with the under manager Jimmy Cregg. At the shaft bottom he said, 'Reight lad, I will leave you now, the loco-man will take you to where you are going'. I had to see Tommy Birk, the fireman, so I got on the loco and rode in. So, for being a bad lad and coming in late, I was put with an old guy called Herbert on a conveyor belt end. I was on there for about a week under where the coal came off the belt, just cleaning up underneath – a horrible job.

The first real district I was on was called the 'North Face' which was down the Dip Side Workings towards the old Calder Colliery. This was quite a steep incline, which went down about 400–500 yards through some air doors to the return airway for the 'North Face'. Through here was a massive tunnel with a rope haulage going in an eastwards direction, inclined upwards to a winch. The tackle used to come through from the main road, through the air doors. We would then come down with our bogies and load them with the tackle. At the winch, we changed direction and went up to the 'North Face' itself. This was a dilapidated old roadway, you could not put your head up – rocks were poking out and the steel plates behind the arches were crumpled and sticking out just ready to take an eye or two out. In fact it very nearly took my eye out one day as we were riding out, a tin plate sticking out caught me and cut my eye – but that was the only time I was injured. They could have back ripped this roadway, but they didn't bother, and being the return airway it was dusty. We would do two trips a day up there, taking two lots of tackle up. At the coalface we would unload and place the tackle at the side of the roadway ready for the men to use. We could then ride out on the bogies. There were about five or six of us tackle lads on this job. At the bottom all the rest jumped off. Me being a bit 'green' I kept on the bogie and around the corner it went up into the air on the rope and me with it. I must have provided a bit of a laugh, because I was then thrown up into the air and landed unceremoniously in a heap in the floor. One time after we had just done a second run up to the 'North Face' and finished, the other lads said to me 'Follow us'. They rolled under the heading, which was about 16 to 18in high – it was so small in fact that the colliers were using the bottom belt of the conveyor belt to load the coal, and the other lads got on this and rode along the face. I just looked on in disbelief, I was terrified, but in the end I followed them. There was a lot of Russians and Poles at Huncoat, for whatever reason I never did find out. On another occasion one of the colliers on the face just rolled over and died on the face; they had to stop the belts and get him out of the pit. We had to place him on the bogie to get him out – the face was stopped all day after that. It was tradition; if anyone got killed or died in the pit all the men would come out as a show of respect.

By 1964 I was attending the Burnley Mining College doing my exams for an apprenticeship for an electrician and passed these. Then I went back to Huncoat as an apprentice electrician, spending most of my time on the pit top, because I was not yet eighteen. The head of the electrical department was Arthur Gibson. We would repair the damaged cables, which had come out of the pit, and put new ends on them and sending them back down. Arthur

Gibson would come every day into the electricians' shop at dinnertime and say, 'Reight, whose going for mi' sandwiches today?' I would usually volunteer and he would give me a shilling and tell me to get him a ham sandwich. One day he came and gave me 2s so I though he must want two sandwiches today. So I got them and took them up to his office and he asked where his shilling change was. 'I thought you wanted two sandwiches with giving me two shillings,' I said. 'Thi' daft bugger, I had no change. If I had given you ten shillings would you have gotten me ten sandwiches?' he said. 'Get out o' office, you dozy sod', he concluded. So I was in the bad books for about three weeks after that. By late 1966 I was underground doing electrical work, moving the panels forward as the coalface advanced, and moving the winches up and repairing signalling equipment and such things – this was on the three shift system. By this time, the pit was coming to the end of its life. A meeting was called in the canteen, and the Union man, Joe Gormley, was there, vowing to do everything possible to save the pit. But the men were not having it – they were all fed up and voted for the closure of the pit. They were convinced that the new manager, Anderton, who came to the pit about eighteen months before, had come there for the sole intention of closing the pit. This was in spite of the fact that a new shearer had just been installed which was all computerised – the first of its type in the country, it could even sense when the weight was coming on the roof and adjust the height of the disk to compensate for this. So, without the backing of the men and the Coal Board saying that they had done everything possible, it was inevitable that the pit was closed. To cap it all I had been diagnosed as being diabetic by this time.

Most of the men were transferred to other local collieries, I had to go and see Bob Watmore at Bank Hall, the manager there, and ask for a job. 'And who are you?' he asked. So I told him, saying that I was an electrical apprentice. 'Ah, yes, I have heard about you' said Watmore, 'You are no bloody good to me, a diabetic, but I have to take you I suppose.' I was sent over to the foreman electrician who told me I would be working on the pit top. Because I was diabetic I was not allowed to go up ladders or handle any moving machinery, so I was put on repairing signalling equipment. This was done in an old tin shack on the surface, open at one end and freezing in winter times with no heating whatsoever. I got out of there by doing other little jobs on the surface; we repaired all the lighting in the new screens, and got stuck into the damaged cables that came out of the pit and put them on drums ready to go to Walkden to be repaired. Even so, I was getting fed up of being passed around doing menial jobs, and in 1969 I decided to leave the pits altogether – two years later Bank Hall Colliery was closed down. After this I went on plant maintenance at a works in Accrington, and was there for two years before I got 'sacked' for refusing to work unpaid overtime.

After a few other jobs, including some time spent working at Nori Brickworks, Accrington, I got a job with the Corporation ending up running the electrical department there before retiring in 1997 after doing twenty-six years' service with them. I enjoyed mining, and all the time I spent working in the pits – even now I still have a great interest in anything to do with mining. I collect commemorative plates about the industry, oil lamps and things, in fact anything to do with coalmining, books and what have you. Another interest in my retirement is quizzes, including the local pub quiz nights at the Crown and the Grey Horse in Accrington.

Moorfield and Whinney Hill Collieries

The Whinney Hill Colliery was located on the right-hand side of Whinney Hill Road at Clayton-le-Moors going over to Huncoat. The colliery dated from 1871, when the *Accrington Times* informed its readers that 'The new pit at Whinney Hill is in full work. Coal and slack may be obtained at reasonable prices from Jonathan Foulds, 1 Abbey Street, next to J.E. Edwards Printers.' Whinney Hill Pit was really an extension of the Moorfield Colliery at Clayton-le-Moors and not that far away. Moorfield Pit only had one shaft, and one of the shafts at Whinney Hill was utilised as an upcast shaft – the other shaft at the Whinney Hill Pit was downcast. It was at the Whinney Hill Colliery that practically all the sixty-eight victims were brought out of the pit following the explosion at Moorfield Colliery in November 1883. At the time of writing, a memorial to this disaster is being erected near to the site of the old Moorfield Colliery on Burnley Road, Clayton-le-Moors. Let us hope that this fares better than the plaque which was fixed to stonework of the nearby Pilkington's Bridge to commemorate the 110th anniversary of the disaster in October 1993: it was stolen by vandals!

Moorfield Pit's other claim to fame is that Eric Morecambe worked there as a Bevin Boy during the war.

The Moorfield Colliery was abandoned in 1948, although the extensive coke works on the site remained until the late 1960s. Today all that remains of the old Moorfield Colliery and its sad past is the capped mine shaft besides the Leeds & Liverpool Canal.

The Whinney Hill Colliery continued until its closure in 1932, although the shafts here were retained for ventilation at Moorfield Colliery until it too closed down in 1948 very soon after nationalisation of the industry. The shafts at Whinney Hill were later filled in and capped off. Directly above the entrance is a small fenced-off area with trees planted directly over the old shafts – this is all that remains of the old Whinney Hill Colliery today.

East Lancashire Mining Memories

The new memorial to the Moorfield Colliery Disaster located near the entrance to the former pit on Burnley Road, Clayton-le-Moors. (Jack Nadin)

Josh Greenwood
Age at interview   *92*
Years in mining   *39*
Collieries worked at   *Whinney Hill Colliery, Clayton-le-Moors,*
                                      Moorfield Colliery, Nabb Colliery, Water
                                      Village, Rossendale, Calder Colliery,
                                      Simonstone, Huncoat Colliery, Huncoat,
                                      near Accrington, Hapton Valley Colliery,
                                      Burnley

I was told about Josh Greenwood by Ian Carter (see Huncoat Colliery) and was able to introduce myself to him at his home in Clayton-le-Moors, where he agreed to put forward some of his reminiscences of his coalmining days in East Lancashire. Josh and his forebears have a long and proud connection with coalmining. His great-grandfather

Edward Greenwood, born in Haslingden about 1835, was a coal miner and a farmer of 8 acres in the 1880s when he was living at 20 Cocker Brook, Oswaldtwistle. Josh's grandfather, Edward's eldest son, James Greenwood, began work at the Broadfield Colliery in the Broadfield area of Oswaldtwistle at the age of just thirteen, in 1870, drawing coal for the colliers. He was eighty-one years old when he finished at the pits; he was underground until he was seventy, and then he did eleven years on the pit top at Huncoat Colliery. Not an easy job either; he was emptying tubs of shale and muck. In the last few years though he got a cabin mending coal sacks and putting tallies on the tubs. Josh's father, also called James, worked at the Scaitcliffe, Wood Nook and Cat Hole (Broad Oak) collieries at Accrington. Josh can also remember that during the 1926 Coal Strike, at the age of eleven, he went with his father to a small wood at Cocker Brook at Oswaldtwistle where they found some stone steps going down into a hole in the ground which led to a small 16in coal seam. He remembers seeing men carrying sacks of coal up the steps to the surface. Here Josh takes up the story of a lifetime spent toiling in the coal mines of East Lancashire.

Josh Greenwood, seen here at his home at Clayton-le-Moors, is still a sprightly gentleman for his ninety-two years. (Jack Nadin)

I WAS BORN at Oswaldtwistle at what used to be the Gardeners Arms pub; this later became a private house and later still the Straits club as it is today. Well, that is where I was born on 26 July 1915 when it was my grandma's house. I had my ninetieth birthday celebrations there a couple of years ago. My mother had a house just below, but my father was in the army, so at that time I was living with my grandma. I had three brothers and two sisters. I had my early schooling at Green Haworth and then at St Andrew's School and St Mary's in Accrington. They had a reunion a short while back for the former Green Haworth boys and girls, but I did not get up. This was a pity really, because I was there when Miss Bunkall was there, a real Lancashire woman, in the 1920s. I left there in 1926 or 1927 and finished up at St Mary's. I had a happy childhood, but it was hard going; my parents were hard-up, my dad was in the army of course. I remember him coming home from the war in 1918 on crutches; I would only have been about three years old then. He had been fighting in France. I left school aged fourteen in 1929 on a Friday, and on the following Monday morning started at the Whinney Hill Colliery. The job was waiting for me, because my older brother John was working there. That is how it was then; if your father worked at the pit or your brothers or uncle, you were more or less guaranteed a job at the pit whether you liked it or not. There was no showers then of course, no canteen at the pit, just the two wooden headgear and the odd surface building on the top, and we would all have to go home in our pit muck. As I left the cage at the shaft bottom my first reaction was that anyone coming down here deserves a wage before he even starts. There was a furnace at the bottom for the ventilation, I think; it stunk of oil and grease. I was told to go through some air doors with my brother. Around my neck was a dog collar from which we had to hang our oil lamps. It was a long way in, the tunnels were only about 3ft high, 4ft at best, and we had to tram in on little bogies pushing on the sleepers with our feet. Eventually we arrived at the main ginney road, then we threw our sledge to one side and went on again – this was the main rope haulage road, the rope ran along the bottom of the tubs. They would couple four tubs up together and send them out to the surface. Most of the coal went to the famed Nori Brickworks and some to the coke works at Moorfield Pit. I was put on drawing coal for the colliers. We had to work in our bare feet, to get them hard for drawing coal, and we would get segs on our feet. It was only in later years, just before the pit closed down, that the lads used to wear shoes, but even then they would cut the back end out of the shoes so that their heel would be bare because their shoe heels would snag on the rails. I was drawing at Whinney Hill until I was about sixteen or seventeen, and then they would put me on bits of other jobs.

The district where the great disaster took place at Moorfield Colliery in November 1883 in which sixty-eight men and boys were killed had long since been abandoned of course. But many of those who perished that day were taken out by the Whinney Hill Pit, the two pits being connected underground even then, and many a time my thoughts would go back to that terrible day as I walked down the tunnels in Whinney Hill Pit. A great many of those who died were laid out in the Greyhound pub just down the road from Whinney Hill Colliery. Underground at Whinney Hill you could get through to Moorfield, Scaitcliffe, Huncoat, Martholme and Calder pits.

The Whinney Hill Colliery was closed down on 2 June 1932. One or two of the lads were kept on to do some salvage work, but I was transferred to Moorfield Colliery. I was only seventeen when I went to Moorfield Pit in 1932 and I was put back on drawing. Moorfield was a wet pit; the only shaft was right besides the Leeds & Liverpool Canal. Whinney Hill though was dry, very dry; we hardly saw any water there. The coal workings at Moorfield were longwall, but all hand got, with pick and shovel, no machines were employed at the face. Unusually, the coal was cut into by the collier at the top of the seam, this was then drill fired down by the shotfirer. A 30in seam was considered good at Moorfield. There was no shift work either at Moorfield, everything was done on the day shift. I was living at Accrington at this time and had to catch the tram each morning in order to get to the pit for the six-thirty start – and we would be out of the pit by two o'clock.

I was once injured at Moorfield with a runaway tub, it was empty, but even so I could not get out of the way. It was the last thing of the day, and we were all camping at this junction and it runaway and I could not get out of the road. The chap at the top was throwing coal in to the tub and it jumped up on to the rails and came down on to me and caught my leg. I was off work for a long time with it, it did not break it, but it was bad for months afterwards. I can also remember a deputy getting killed at Huncoat; he was piecing up a face conveyor belt when the drive broke loose from the sprags and crushed him to death against some steel props.

I was only at Moorfield until 1938, during this time I had met my wife Ellen Bromley through dancing, she was a Crawshawbooth lass from the Rossendale Valley, and we were wed in 1937. We were wed fifty-four happy years till she passed away in 1991.

After this I went to live at Palace Row at Crawshawbooth in the Rossendale Valley and managed to find a job at the Nabb Colliery at Water village. I can remember Stephen Spencer [see Nabb Colliery] and his dad, Stephen, and also Jimmy Cropper at Nabb Pit. Jimmy Cropper opened a pit of his own after; he had the pick of the men after Nabb had closed, he got all his tubs,

A group of miners from the Burnley and Accrington area at the Miners' Home at Blackpool around 1934. Josh Greenwood is second from right in the back row. (Josh Greenwood)

rails and sleepers from the old Nabb Pit. Jimmy was a farmer, he had three brothers in his pit, and his dad was the deputy. I liked it at Nabb Pit, they were a good set of lads to work with, they all worked with their shirts off with just candles for light. It was also easier drawing, because the tubs were smaller than anywhere else – but I was dattaling most of the time. Billy Holding was the under manager, he was a nice chap.

While I was at the Nabb Pit there was a strange incident in late August 1939. Two thirteen-year-old lads, Jack Hitchen of Spring Garden Cottages and Fred Williams of Back Carr Road at Water village, had set off over the moors saying to their parents that they were going to look for an air-raid shelter. That was the last anyone ever saw of the two young lads. It was thought that they had gone into some old mine workings on the moors, but theses moors were littered with old workings. It was thought that the two lads had gone into the old Dean Colliery, which had been abandoned for decades, but which linked up underground with a number of the other local pits such as Nabb and Grimebridge. The alarm was raised, and while scores of anxious relatives and volunteers scoured the open moors, miners from Grimebridge and Nabb pits, including myself, set about looking in the old workings. I remember seeing the fireman's chalked marks on some of the old roadways down the abandoned pits – some dating back to the First World War. The boys were never found, even today it remains a mystery in the Rossendale Valley just where these lads got to – and will they ever be found?

It was while I was working at the Nabb Colliery that I got my shotfiring and deputy papers by gong to night school at Burnley College. There was no chance of using my papers at Nabb Pit though, it was all a family affair. The only way I would ever get to use my papers at Nabb Pit was if someone died. So after about four or five years I managed to get a transfer to the Calder Colliery near Simonstone, between Padiham and Clayton-le-Moors. There was always a lot of gas at Calder Pit, a chap died there, he was gassed. They sent him from one district to another for a spare battery, and after he had been missing for a while they went looking for him. For some reason he had gone into some old workings, and the gas must have just hit him and killed him. Another fellow, George Thurston, a shotfirer, was also overcome with gas, they had to bring him out to the surface to revive him. He finished up at Thorney Bank Pit.

I came out of the pit in 1940-something, and went to Cortould's on the engineering staff, but then Cortould's finished and went to Preston – I could have gone to Preston with them, but I went back to the pit. I was at Calder Pit almost until it had finished in July 1958, when I was transferred to Huncoat Colliery; Jimmy Cregg was the under manager here. Ralph Taylor I remember was a deputy at Huncoat. At Huncoat I was put on ripping. I still had my fireman's papers, but did not renew them – anyway tunnel ripping was more money. We could earn about £27 a week doing this, good money, really good money, for that time. I can remember a couple of accidents at Huncoat while I was there. One day we were riding out on the conveyor belt when someone shouted out 'Stop the belt!' But it was too late, the man at the end had been resting on the guard fence which gave way – the belt caught hold of him and took his hand and arm clean off at the elbow. I can still see him now, holding his arm trying to stop the blood. On another occasion a young lad working on the coalface was taking some props off the scraper conveyor. He took the prop off at the front, something we were all told not to do. As he lifted the prop off, it raised itself up towards the roof, trapped his hand and took with it two fingers.

Huncoat Colliery had undergone redevelopment then, a huge amount of money was going into the pit. Diesel locomotives were introduced at the shaft bottom to haul the coal onto the main levels. The Upper Mountain Mine had also been developed, and surface reorganisation had taken place. They were driving a drift up underground, at about 1ft in 3ft, to connect with another seam; they had the foreigners on this job, the Irish, French, Poles, they came with a mining company. These foreigners were really hard workers, they could really graft. I was on the day shift at Huncoat when Hapton Valley went up. We knew about it in minutes. What a bad do that was;

we were all stunned. Some of the firefighting team at Huncoat went over to see what they could do – well we know all the rest don't we?

In spite of all this investment, the Huncoat Pit was closed down in February 1968; the men even voted for it to close down. I was at the meeting when Joe Gormley was there trying to save the pit – but we all knew that any coal left at Huncoat was all under the housing and industrial estates and the Coal Board would not mine under those, it would cost too much in subsidence compensation. After Huncoat closed I was transferred yet again to Hapton Valley – I was beginning to get a bit paranoid, every pit I had worked in had closed down. At Hapton Valley I finished up on the console in the main gate, a bank of control and telephone panels which controlled the scraper, belt and things like that. This was on the No.5 District at Hapton Valley – they had a big roof fall here later on. They were putting a return wheel in for the haulage and took away some of the floor to set the legs, but they started sinking with the weight on the roof, and the whole lot came in. I found conditions at Hapton Valley Pit very hard and wet, on one district as many as seven pumps had to be running full time to avoid flooding the coalfaces there. I worked at Hapton Valley Colliery until June 1979, the pit closed down in 1982, and I finally retired from the industry altogether.

I still enjoy a game of bingo twice a week at an old folks' place on Sparth Road, and I look forward to it, it is a social outing for me as well, with cups of tea and biscuits. Any birthdays or anything like that we all get together and have a party, and all enjoy ourselves. I also like a bet on the horses, even win sometimes.

11

Nabb Colliery

The Nabb Colliery was one of the many little drift mines or tunnels driven into the hillside which worked around the bleak Rossendale moors. The colliery was located off Dean Lane at the village of Water to the north-east of Rawtenstall, although on a number of documents it is listed as being at Dean, the small former cotton-weaving hamlet near here. The Nabb Pit dated from around the 1860s, being one of George Hargreaves' pits – this firm owned a number of the Rossendale pits. Nabb Pit was always regarded as a family pit, where brother worked alongside brother and father and uncle, generation after generation. A tales is told of two men, father and son, both of whom had a terrible stammer, who were having an altercation at the pit. The men coming out of the pit at the end of the shift saw the lamp attendant hanging over the barn-type door of the lamp room laughing hysterically. 'What's up?' asked one of the colliers. 'Well,' said the lamp attendant, 'I have been here half an hour, and they have not even started arguing yet.' There was always a lighter, more humorous side to mining than the hard work, as this example shows.

Nabb Colliery, like many other of the little Rossendale pits such as Old Meadows at Bacup, always worked on the old-fashioned principle of pick and shovel, with outdated chain haulage systems. But nevertheless through the hard work of its labour force these little pits made healthy profits. The little Nabb Pit however did not fit into the new National Coal Board's image of its 'new modern mining patterns' and soon after the 1,500 pits in Britain were nationalised in January 1947 faceless bureaucrats began to probe into the future of Nabb Colliery and others like it. It was decided to close the mine down. It must have been a sad day in March 1954 when the twenty-seven men still remaining at the pit walked away for the last time. In its last year of full production, though, the little Nabb Colliery raised 5,124 tons of coal. In 2005 an application was made to convert the brick-built former pithead baths and the block-work boiler house into a dwelling house – but this was refused.

∞

Stephen Spencer
Age at interview   80
Years in mining   20
Collieries worked at   Nabb Colliery, Rossendale, Huncoat Colliery, Huncoat, near Accrington, Gambleside Colliery, Crawshawbooth, Rossendale

I meet Stephen Spencer at his home in Rawtenstall, and was very pleased that I did. I never expected to be able to interview anyone who worked at the old Nabb Pit, which closed over fifty years ago. He is also probably the only living person who went down the old Gambleside Colliery – and one who was able to bring out a piece of coal from there. Stephen still takes an interest in anything to do with coalmining, and was also able to put me in touch with other Rossendale miners. Thank you, Stephen.

STEPHEN WAS BORN on 29 September 1926 at 903 Burnley Road, Crawshawbooth, between Burnley and Rawtenstall, the eldest child of Stephen Spencer, a collier at the old Gambleside Colliery and mother Katie, a weaver. Stephen's birth came at the end of the 1926 Coal Strike in which his father was involved, and consequently times were hard. However looking back Stephen says that everyone else was in the 'same boat' and that they did not see it as poverty, it was just an accepted way of life. A number of other neighbours worked at the Goodshaw Hill Colliery on the other side of the valley from Gambleside Pit. Even when the 1926 strike was over, they were on short time; Stephen's father and the others might just work one day a week. Often they would be laid off for a week at a time. In 1936 the family moved from Burnley Road to Swinshaw Farm Cottages, besides the farm at Loveclough. Stephen and the other lads loved it around here 'The moors as far as we could see were our adventure playground, and once on the moors, the parents did not see us again till teatime' he said. Rain or shine, schooldays started for Stephen and his younger brothers with the tramp down from Swinshaw Farm Cottages to the main road to catch the bus to St James' School at Rawtenstall every weekday. It was the same going back; they got a ticket off the headmaster for the bus to take them back – this was in all weathers and nobody complained. When he started at St James' 'little' School, he was four-and-a-half years old, and then he went to the 'big school' next door.

Stephen Spencer at his home in Rawtenstall in September 2006. I am extremely grateful to Stephen as he was able to put me in touch with a number of other Rossendale Valley coal miners – adding much more to our knowledge of mining history in this area. (Jack Nadin)

Stephen recalled the day he set off over to the Gambleside Colliery to meet his dad and take him his bait when Jim Harry Ashworth, the colliery under manager, saw him sat in a field. He said to Stephen, 'What thy doing here, young man?' Stephen explained, 'I'm waiting fur mi' dad'. So the manager took Stephen to the pit-top cabin. Stephen remembered walking round the pit top and looking down the shaft, where his dad was working hundreds of feet below. When his dad came out and repeated the manager's words, 'What t' doing here?', Stephen explained that he had brought his bait up for him.

His dad had his bait, and then the manager said to his dad, 'Dusta think tha' can look after him?' and his dad said, 'Aye, I'll take him back in with mi'. So the young Stephen was taken into the pit and watched his father knocking props and getting coal till the end of his shift and was brought out. The young lad remembered getting a piece of Gambleside coal before coming out with his dad. That, said Stephen, was his first taste of mining. He took the piece of coal home to his mother, and being a bit dirty from being down the pit, his mother did not know whether to be mad or not from not knowing where he had been, or to be pleased to see him. Soon after this the Gambleside Colliery was closed down: its main customer was the Sunnyside Printworks in the valley below, and when this closed, so did the pit. On the closure of the old Gambleside Colliery, Stephen's father was transferred to Nabb Pit over the other side of the hill.

Young Stephen left school at the age of fourteen when they were living at Collinge Farm to the west of Leebrook Road at Rawtenstall. On his last day but one at school he went home and said to his mum, who was busy in the kitchen making the tea, 'They want to know when I go back to school where I am going to be working'. His dad was sat in a chair reading the newspaper, and without even looking over or rustling the paper said to young Stephen, 'When t' goes back to school, tell teacher that thi' going down pit wi' me, I've had your name down for three months'. On the Monday following, which was his fourteenth birthday, Stephen was at the pit with his dad. He had everything ready, including his tram and his locker. They went to Nabb Pit in his father's car, which was parked in the pit yard on Dean Lane. From here they had to walk up Dean Lane to go through the farmyard at Springhill Farm and then walk across till they met the surface chain road at the pit and then turn up to the pit top. In summer this was grand, but in winter not so good. Stephen recalled one really bad winter's day which nearly put paid to the under manager at the pit. The snow was really coming down, and several of the workers were in the cabin at the sett end. Stephen's dad said to him and another miner called Rowland Broxton, 'Right we are setting off, we'll stick together', so they set off up the surface chain haulage track. On the way up, young Stephen and Rowland came across a tub tipped up on its side, so they stopped there and got inside to have a rest. They set off again, and within a few yards came to a snow drift about 6ft high. They finally arrived at the pit top – Stephen's dad had already passed them, but they did not know it then. The men then realised that Billy Howorth, the under manager, was missing – it was known that he had set off but he had not arrived. So a number of men set off in search of him. He was found about 50 yards away from the surface chain track sheltering behind a hen hut. He never worked again at

Nabb; after this the under manager was Jesse Rawstron. On a number of occasions the men at the pit would have to come out of the mine and were put to work clearing the snow off the surface haulage track – but the day after it would be just as bad!

Stephen remembered his first day at Nabb Pit. His dad had his tram ready, and he was given four candles by his dad at the pithead baths steps. [Stephen smiles here, recalling the Ronnie Barker TV sketch about four candles – or was it fork handles?] 'Dad took me in the pit [and we went] about 50 yards straight on, and then we turned right about 55ft and then you went along a brick-arched roadway; it was level, but in winter times, it was that wet it just poured through, because there was so little cover over the workings. Your feet were wet through going through here, even though the rails were lifted off the floor a foot or so. Then you got to a roadway propped by bars and wooden props and you went on from here down a brew until it levelled out until you were about 400–500 yards into the workings, to the fireman's cabin.' Young Stephen was going up the No.2 District, up to the left-hand side; his dad was working in the No.3 District to the right. Stephen was put under a collier named Walt Fletcher, who took him up to the far end. 'I had to leave my dad!' Stephen exclaimed as he recalled that day. He followed Walt, still on their trams, for another 400–500 yards to where he was shown the 'bait oil' and here he was introduced to the other lads. 'They all sat round, them looking at me, and me looking at them, with our candles,' said Stephen. There were no safety helmets at Nabb Pit, everyone wore flat caps, and the drawers had a pad to protect their heads while pushing the tubs. To start off, Stephen was placed with two colliers, Squire Rawstron and Sammy Fletcher, near the ginney head. This meant that he only had a short distance to draw the coals – 'but after this you went in with the rest of the drawers'. The tubs they used at Nabb Pit were made of steel and would hold 3½cwt, and the tub itself weighed about ½cwt. 'I'd a job to shift them,' said Stephen. 'The ginney tenter, a lad called Mark Heys, had to come and give me a lift – don't forget I was only fourteen. But don't forget either that the other drawers were the same age. Rowland Broxton had started about a fortnight before me. The other drawers were Walt Fletcher, and Jack Nuttall.'

Going out of the pit, you got to a place called 'Clifton Flats' where everyone waited until their time was up, but Stephen was allowed to go out of the pit early in his first week to get showered and to get used to the procedure, but after that he had to 'muck in with the rest'. The pay at Nabb was based on work done from Wednesday to Wednesday, and they worked Saturday mornings. Stephen's first week's wage was 17s 6d:

To get paid you had to go to the sett bottom where your money was in a big tray in little squares and handed out. Then you had to pay the Union man; he was in what was like a sentry box in a corner of the sett end. The Union man had to get permission off the mine owners, George Hargreaves Collieries Ltd, to put this box there.

Drawing coal, it was steep, shoving and pushing, but the Lower Mountain Mine was not as high, and you got some sore backs. At best the Lower Mountain Mine was 3ft high, then you could take 4 or 5in off that for the roof timbers, and then another 3 or 4in for the rails and sleepers and you are down to about 26in. The drawers were always catching their backs on the timber roof bars, which made them bleed, and then with the next tubs out, you would catch them again – it were painful! Then the blood would stick to your shirt, and in the baths you had to pull off your shirt trying not to break the scabs – but you always did. You tried not to whinge or cry in case others were watching you, and the chorine in the showers stung, and then the next day it would all start over again, catching the scabs on the roof.

It was while Stephen and Maurice Pickup were working as drawers for six colliers, Tommy Parkinson, Stanley Binns, Dick Cropper, Alf Chadwick and two others, that there was a dangerous incident at the pit. Maurice said to Stephen, 'I think it is weighting a bit where the colliers are working.' Stephen tipped his tub up and put it on one side and waited for Maurice to come down, and noticed that the roof was also weighting where he was. He shouted up to the colliers, who were about 20 yards away, to warn them. Then the props started creaking and the roof bars started splitting, and the coal started bursting off the sides of the tunnels. Everyone was now aware of the danger, and they all set off down the tunnel. It seemed that the whole pit was on the move. They all made it safely down to the ginney head when the whole lot came in with a loud rumble. When they looked back at where they had all been working a few minutes earlier, it was like someone had built a stone wall – the lot had gone. One of the lads, Ab Chadwick, had stumbled on his way out, and George Cutting grabbed him and dragged him out. When they got to the end all his shins were bleeding where he had been dragged out over the sleepers. They had a head count and one person was still missing. 'I thought it was Tommy Parkinson,' said Stephen, 'but he did turn up, he had gone another way round. That was talk o' pit for weeks'.

Nabb Pit was always a family pit where son followed his father, and brothers worked alongside each other. There were the Broxtons: Jimmy Broxton was the dad and Stanley Broxton his lad, brothers Jack and Roland Broxton; there was Roland Dunn, Ralph Dunn and nephew Jack Dunn, then there was Billy Baldwin, Bobby Baldwin; Sam Fletcher, and his two lads, Walter and Jack; then Harvey Smith and Freddy Smith; Freddy Rushton and Dickie Rushton; Jimmy Cropper and Dick Cropper, and so it went on. Of course you would have your differences and arguments between each other – but once they were sorted out there was no malice, you forgot all about [it] and got on with the job.

Stephen was on drawing for a number of years, and as new lads came and colliers retired, he got put on datalling, laying rails and other odd jobs. However one day the manager said to him, 'Get your tackle together, you are going coaling'. The union sold the tackle to the colliers at this time, and so Stephen went and got two pick shafts, four pick-head blades and a 7lb hammer and a little box you sat on while getting the coal.

There was no training given, you got your tackle, went into the pit, and they said you are working there! Knowledge was expected to be learned while you were drawing coal, through observation of the colliers at work, and learning from them. There was never any machinery installed at Nabb Pit, it was all hand pick and shovel. The only bit of mechanical apparatus used at Nabb Pit was a hand drill on a steel tube which was wedged between the roof and floor, this had a handle on which had to be cranked to drill into the coal, after which the holes were then stemmed and fired down by the shotfirer. A period of probation followed on the coalface, just to see if you were actually capable of doing colliers' work – but how you ached those first few days. You had segs on your hands you could have struck matches on.

Wedges were used to pull 'cobs' off the coalface and you only had about 7 or 8in over the top of the tub to get it in. It used to pull on your guts as you tried to turn around and get this piece into the tubs. You also had to shovel the smaller coals though this small space. The Lower Mountain Mine at the pit was only about a yard high, sometimes under that. The Union Mine, where the Lower Mountain Mine met the Union Mine, could be up to 5ft high.

Nabb was an old pit, and Stephen thought that the workings here went over as far as Weir Terrace, about three quarters of a mile away. Water collected in a district called 'Earby Clough' in the lower dip workings. Stephen was once sent down Earby Clough while on probation and while working here his pick went though and broke into some old workings. By this time, the workers at Nabb had progressed from candles to carbide lamps, and he shone his light into the old workings – there were no timbers up, and the width was about 15ft wide. As Stephen went forward, he had to timber up for safety's sake through to the next lot of coal, which once removed, revealed another open old working. The coal had been left by the old-timers to support the roof – but no one knew about these workings, they had never been plotted or mapped out. These old workings might have contained blackdamp, a suffocating gas. Methane, or firedamp, was never present at Nabb Pit, and this is why the men were allowed to use candles and carbide lamps, even to smoke in the workings. Eventually a fan had to be installed on the pit top at Nabb Colliery because the air got a 'bit stale' in those workings which were in the dip of the seam where the blackdamp accumulated. Prior to this the pit was ventilated naturally. 'Candles would not burn, and the damp caused fierce terrible headaches among the men and boys,' Stephen recollected.

> Nabb Pit had a wonderful method of allocating workplaces in order not to cause friction or discrimination among the men. All their numbers were placed in a box, and the men drew a number out, rather like a raffle, and that number was their workplace for the week. The management could not therefore be accused of sending a man to a bad place to work, and others to a good place to work – simple but effective. If you did happen to have a good place to work though, and you could fill more tubs because of this, the management might come round and take a halfpenny or a penny off the pay per tub. While this did even things up a bit for those working in the harder place, it also meant that those in the easier places had to fill more tubs with coal to get the same money.
>
> Your bait consisted mainly of jam and bread, especially me, because my dad once got a tub of apricot jam which must have weighed half a hundred weight, and we had jam for months after. To drink we used to take cold tea in a metal bottle – cold tea is a great thirst-quencher. We could warm it up a little if we wanted, by putting the bottle on one side, at an angle, and putting our candle underneath – it would warm it. The candles come in for a lot of things, lighting your fags, and if you crouched down, not forgetting we were often wet through, and put your jacket over your head, it would warm you up.

Stephen was able to recall an amusing incident at the pit. One day Jimmy Broxton was having his bait along with the others, all sat round in a group, when Stephen Spencer senior, Stephen's dad, asked, 'What have you got for your bait Jimmy?' Jimmy, lifting the 'lid' on his butties, replied, 'It looks like bit of boiled ham, Stephen'. 'Eh, that's posh,' said Stephen's dad. 'Aye, but it's a bit dry, not reight good,' replied Jimmy, and when he had a closer look at the filling it was a piece of blotting paper. 'No wonder it tasted rotten,' said Jimmy. The lads had switched his ham for blotting paper, and when he looked to see what was on the other butty, it was the playing card the Ace of Spades!

'There were no major accidents at Nabb, we got the odd busted finger and that, but nothing serious.' Stephen recalled the time when his father broke his ankle at Gambleside Pit, and they brought him out of the pit and took him home on a bicycle – all the way down that rough track to the village on a bike with a broken ankle! While he was off work he got £10 a week on the sick. He got so desperate that he told the manager that if someone would help him to get to the pit top, down the pit and on to his tram he would manage to get to his stall and he would be able to get coal. Of course the management refused.

> We found out that Nabb Pit was going to close, although we had not got a date yet, and a man called Ronnie Smith used to come round dust sampling, and he said to me, 'Do you fancy this job?' I said, 'Aye, pit's finishing, there's no future here'. So Ronnie said he would make some enquiries, and got me set on.

Stephen and a couple of lads called Archie Harrison and Jack Howe then became dust samplers at the eighteen pits that were then working in the Burnley area, a job he did from 1952 to 1954. It was a job with 'wark took out of it though', but it gave Stephen the chance to look at the other pits in the area, and he fancied Huncoat Colliery at Huncoat near Accrington. He went on to work here for eight years as a shotfirer.

One day while in the showers he noticed that he was spitting blood. Not good, he thought, and went straight to the doctors, who signed him on the sick. While he was still on the sick, he managed to get a job as an ambulance driver at Stacksteads, a job he did for thirty years until he retired in 1992. 'It was a rough life in the pit,' said Stephen, 'it was not easy by any means. But it made you appreciate other things in life, and it was an industry I was glad to have been part of and I have no regrets about it.'

Red Earth Drift Mine and Pickup Bank

A drift mine worked by Shaw's Glazed Brick Co., Red Earth was situated in Yate and Pickup Bank was near Hoddlesden, Darwen. Red Earth was mainly a fireclay works, although coal was extracted from the Lower Mountain Mine and used to fire the kilns at the works. The mine was William Henry Shaw's brick company's third drift mine, the others being Waterside and New Waterside Drifts, and opened around 1938. Its closure in 1963 marked the end of coal and fireclay extraction in the Darwen area. However, the glazed brickworks are still in operation, although its output is now confined to the manufacture of 'Belfast sinks' along with specialised bricks and the like.

∞

Bert Holdsworth
Age at interview   71
Years in mining   9
Collieries worked at   *Hoddlesden Colliery, Hoddlesden, near Darwen and Red Earth Drift, Waterside near Darwen*

I have but the fewest details of the Red Earth Drift Mine at Waterside near Darwen, and knew little, therefore, of the working conditions, the height of the seams, or the men who actually worked there. I was particularly pleased then when Bert Holdsworth of Balderstone contacted me about his time spent at the Red Earth Drift. I interviewed him at his lovely little cottage in the open countryside near Balderstone Village – where Bert took up the story in his own words.

I WAS BORN at Mellor, north-west of Blackburn in Lancashire, on 28 November 1934. I am a child, one of eight, of tenant farmer Albert

Bert Holdsworth and his wife Olive in their delightful 140-year-old cottage near Balderstone village in Lancashire. (Jack Nadin)

and Ethel Holdsworth. We had a good childhood considering, living in the countryside, but sometimes things were hard. We later moved to Oswaldtwistle, and I left the Ryddings School at Oswaldtwistle at the age of fifteen. My first job was farm labouring, but the money was poor and I was tempted into coalmining after a couple of years, this paid more, £4 instead of £2 – a big increase. I got set on at the Hoddlesden Colliery after the usual eight weeks' training at Bank Hall Colliery at Burnley. We were living at Blackamoor and I got to the pit every day on a pushbike, about two and a half miles there, and two and a half miles back again. As usual for the young lads starting at the pit my first job was that of a drawer, taking coal from the colliers in tubs and pushing them out towards the main haulage underground. It was only a 15in seam at Hoddlesden Pit, and the coal was all hand got, with pick and shovel. The first day down the pit I was put under the care of one of the firemen, Stan Oldham, I think his name was. His only guidance to me was 'Keep your head and your back down lad'. I soon learnt why, I kept catching my back on the roof supports and soon it was covered in scabs – they were sore. Len Aldridge was the manager, and another of the firemen was called Jock Walton.

There were some scary moments at Hoddlesden Pit, especially on Saturday mornings when working the coal back from the boundary and we would have to go in and withdraw the props. We would cut a 'nick' out of the props and pull them out with a 'Gablock'. We would sit and listen as the weight of the roof came down on the props – they would creak and groan before giving way, and then the whole roof would fall in. On one occasion we all got trapped behind a large roof fall, about 30ft of a fall on the main gin line underground. There was about twelve of us, all on the wrong side. We attempted to get out of the pit by going round a back road which was supposed to be inspected by the fireman once a week, but we could not get through. In the end they had to bring in about seven or eight miners to start digging from their side, while we dug from our side, and eventually we managed to get over the top of the fall. It took about ten days to clear all that roof fall, we had to reset all the props and put the shale back over the top. My mother was really worried about me, she was pacing up and down the lobby, she even rang the pit up to see what was the matter, and they just said that we were working overtime. We were trapped a while, about ten hours in all – but you work harder don't you when you are trapped, you just want to get out. We had a Hungarian lad working with us when it happened, Frank Maitz. He is in Scotland now. It was like a big ballroom when the roof came in. To repair the fall we put in a platform up to the roof, and while we were on the platform, me, my brother and this Frank, I saw a great big rock on the side coming in and I saw this hole in the floor and thought I can just about make through there before it comes. Just as I was about to set off, Frank grabbed me by my shirt and the cable on my lamp and pulled me back. No doubt he saved me; if I had gone through the rock would have hit me. We are still in touch with Frank to this day. I also remember the day when another large fall came in at Hoddlesden, which killed two men – but I was at Waterside at that time.

It was hard at Hoddlesden Pit, especially winter times getting up there. I have seen tubs run away over the moors and make a right mess, I have seen six or seven tubs all piled up on top of each other up there. Some of the workers used to ride up to the pit on the tubs, but one fellow got killed one night where the gin went under a bridge – he was trapped because he did not jump off before he came to it.

I was only at Hoddlesden about twelve months, and then I went to Waterside pit – I was about eight years there. The fireman there, Jock Walton, was talking to my mother one day, and he said they were after some more at Waterside, and the pay was £2 more. They would take the coal and about 2ft of the fireclay underneath it, the coal was used to fire the kilns in the yard at the works, and the fireclay of course was used to make the earthenware goods. The tunnel besides the little brook, which can still be seen, went to the bottom of the airshaft – there was a big fan there at the shaft. This tunnel was more of an escape route in case of an emergency. At Waterside I was still on drawing at sixpence a tub, although I did do a bit of dattaling at times. To get into the pit there was a hole about 4ft high and 6ft wide in the floor; we walked down about 300ft to the seam, then three quarters of a mile or so to the bogies. At the face, the coal was got first, and then the area cleaned up to get at the fireclay. This was drilled and then fired down, cleaned out again and the rails were put down to get at the next lot – the stalls were about 6ft wide. The tubs came out of the pit on wire ropes and on the surface they were lashed to another rope, fifteen at a time, to go on to the works about three quarters of a mile away. The surface haulage was worked by a large electric motor – although the haulage engines underground were the usual 'Pickrose' type. Later though they closed this surface gin down and got a large hopper on the surface where big wagons came and backed under and took the clay to the works that way. Eventually the workings at Waterside were 900ft under the Grey Mare pub on the Haslingden Road. Although it was a private pit, we still came under National Coal Boards rules, and the Mines Inspector would come around every now and again. We knew when the Inspector was coming.

One of the worst jobs was stone dusting, and it was a case of getting the props up as well. To get paid, a man on the surface would fill in time sheets and send them over to Shaw's Works where the wages were worked out in the pay office, and the money was then sent over to the pit.

Jacky Beran was one of the characters at Waterside; he would come to work in his wife's silk bloomers. We hardly wore any clothes at all down the pit, any old pants would do – there was little point in wearing a lot of clothes because of the water in the pit, it was often 2ft deep in places.

I enjoyed coalmining, and I preferred Waterside Pit to Hoddlesden. The wages were more than farm labouring, and of course we got Concessionary Coal which was a bit of help in winter time. At Waterside you got paid your normal money up to twenty tubs, and then after that you got an extra sixpence a tub, which boosted your wage up. One ploy the colliers used was to cut up the old pit props and put them in the bottom of the tubs – they could gain about two tubs extra a day like that.

The main drawing ways at Waterside were about 46in high and the tubs came out of the pit in sets of fifteen. We could not use that roadway to get into the pit, we used another one, and we would sledge in on a little bogie – the distance would be about half a mile to the far end. On the surface at

Waterside was a workshop where they mended the tubs, and a cabin where they serviced and handed out the lamps. The tubs, the new ones, came in packs. The tubs themselves were about 4ft by about 2ft high – they would hold about 7cwt of coal, and about 8cwt of clay. On the downhill gradients we would put a 'shoe' on the tubs, which acted as a brake; this was fastened to the hook on the tub with a chain and then placed under the wheel on the tub. Once we got to the main gin road we would lash the tubs onto the rope with a chain.

Waterside was a wet pit, there was about four or five pumps going all day. The fireman who got me the job at Waterside used to work on nights attending to the pumps, and one night he went into the pit and there was a cow there – that must have frightened him when he saw its red eyes from the beam of his lamp. The pumps would work from the far end of the pit pumping to another one, which then pumped to another, and so on till the water was got to the surface. Here, the water was put in a huge tank, and then it was pumped again up to the main works, and used to soften up the clay and things like that – nowt were wasted at Shaw's.

Bert also recalled that on one occasion a bungalow disappeared down one of the old shafts at Hoddlesden Pit.

I could see the piano down the shaft. The owners were away somewhere on holiday – they must have had a shock when they came back.

When I first got married and went down the pit we had a little farm called Hole House near the pit; the powder magazine was in the corner of our meadow – so near. One day the neighbouring farmer came along and said, 'The front wheel of my tractor's gone down a hole, can you bring your tractor and pull me out?' When I went over the wheel had gone down into the old workings at Waterside Pit – it was only the plough on the back end that stopped him going in completely.

When I finished at Shaw's, about two years before the pit closed down altogether, I went working as an agricultural contractor for about thirteen years. Nowadays I spend my time doing a little gardening for the locals, helping out here and there, and keeping the wood-burning stove supplied with logs – that keeps me busy.

Scaitcliffe Colliery

Scaitcliffe Colliery was located at the top of St James Street, Accrington. The first pit was sunk by George Hargreaves & Co. in 1859 to the Upper Mountain Mine at a depth of 180ft. The seam was 30in thick on average, but the mine was abandoned in 1883. A second attempt was made to work the coal seams here during 1890-91, when two 540ft-deep shafts were sunk to the Lower Mountain Mine. A wooden headgear 10ft in diameter was built over the upcast shaft. The upcast shaft was 14ft in diameter and had a single cage, wound by a small steam winder via a pulley built into the engine house brickwork. The pillar and stall method of mining was used in the early days of the pit and coal was hauled from the face in 4cwt tubs by endless chain haulage, whose main engine was situated at the shaft bottom. When it closed on 29 June 1962, Scaitcliffe Colliery employed sixty-two men underground and fourteen on the surface.

∽

Bill Walsh
Age at interview 69
Years in mining 9
Collieries worked at *Bank Hall Colliery, Burnley, Scaitcliffe Colliery, Accrington*

I met Bill Walsh at his comfortable little flat in Accrington near the town centre. He had replied to my plea in the Accrington Observer for mining memories. Here, Bill relives his time spent at the Scaitcliffe Colliery in Accrington, recalling life underground and a remarkable escape from death while working the coal cutter in the 18in-high coalface there – his companion though was not so lucky. Yet another example of the harsh and dangerous conditions the miners worked under in the East Lancashire coal mines.

A group of miners at the old Scaitcliffe Colliery at Accrington. Notice all the flat caps! The group in the front of the photograph appear to have been very young indeed. (Josh Greenwood)

MY PARENTS WERE Jack and Maud Walsh, *née* Furness, and I was born on 25 September 1937 on Cross Street, Black Abbey, Accrington. Jack, my father, was a miller in the metal trade. My mother worked at Grimshaw's stout-bottling works in Jacob Street. At the age of seven years, we moved to Ranger Street. I received my education at St James' School, both junior and senior schools. The head teacher here was a Mr Barton; other teachers included Mr Day and Miss Pickup. I left school at the age of fifteen and started doing welding at Wellock & Bailey's at Church – but I only did this job for twelve months. I applied to go down the pit when I left school, but Mother would not let me go down – she even hid my application papers so that I could not go. However, I found out where she had hid them, it was in some drawers, and pestered her that much that in the end she gave in and let me go. You had to do some compulsory training first, and this had to be done at the Bank Hall Colliery at Burnley. To get to Bank Hall Pit I used to catch the bus at Accrington centre and get off just outside the pit gates on Colne Road in Burnley. It was exciting for me on my first day at the pit, I was not scared or anything like that. I was not even apprehensive about going down the pit shaft in the cage, even though they used to drop the cage in the shaft.

Once training had finished we were given the choice of which pit to go to, one nearest our home, and in my case this was the Scaitcliffe Pit at Accrington. The year was 1953, and I was aged sixteen. To get into Scaitcliffe Pit yard we used to have to go up some steps, about a dozen steps or so. On

Bill Walsh at his flat in Accrington on 5 October 2006. Unfortunately Bill died in March 2007 before this book was published. (Jack Nadin)

the far right-hand side of the yard was the lamp room and a brew cabin for those working on the pit top. The winding house was at the back of us as we got to the top of the steps, and the shaft was in front of that of course. The screens and bunkers were on the far left-hand side – this is where the wagons used to come in and take the coal away. The wages down the pit were huge compared with those I was getting doing welding work; it rose from 25s to £6-odd. As was usual, I was placed on the pit top for six weeks in the screens and taking the tubs of coal out of the cage at the pit top. After this, six of us were allowed to go underground under the care of Charlie Bond and another fireman called Norman Williams – he used to take us sometimes. Another fireman was Joe Astley, he came from Blackburn, and my brother Joe Walsh was a fireman at Scaitcliffe. We were put on what they called 'timber work', taking the props to the miners working on the coalfaces. The colliers worked on the face during the day, and the rippers would come in on the night shift and rip down the roadways. One of the firemen, I will not mention his name, was a right dirty swine, always playing tricks with the new young miners. He would bring a piece of bacon down the pit with him, and stick it up his bum. He would say to the young lads, while he was going along the coalface, 'Here, get my bait, will you?' 'Where is it?' the young miner would say. The fireman would show him his bum, and say, 'Up theer, can you get it out for me?' Most of them were sick, and can you blame them? Another 'prank' was to put a dead mouse on your butties; humour was always a bit sick down the pit. It was always wet at the bottom of the shaft at Scaitcliffe, it was as if the clouds had opened up. The bottom of the shaft was lit up by electric lighting, but just around the shaft area, about 30 yards.

Once we had done our training with the firemen, we could do what we wanted, and what shift we wanted to work on. You could go on to moving the conveyor over, after they had cut the coal, or other face work, or go on to what I did and that was working the coal cutter. They worked longwall faces at Scaitcliffe, undercutting the coal, and then firing it down. The cutters worked a jib, rather like a chainsaw, and was on armoured conveyors, and the cutter was pulled along the coalface by wire ropes, cutting the coal as it went. When the coal had been cut, the men employed on moving the conveyors over came in, moved them over and then reset the props. From the shaft bottom, we would go and get what we called a sledge, which was four wheels and a flat board. There was a flap on the sledge, and we would lift this up and push it under a wire rope haulage and trap the wire with the flap by kneeling on it. This would take us into the pit about a mile. From here, we had to 'pick' our way in, this meant sitting on your sledge and pushing yourself along with your leg behind you pushing on the sleepers. It would take us one-and-a-half hours to get to the coalface, three and a half miles in from where we got off the wire rope – and it was all uphill. The tunnels here were 32in high at best, most of the time a lot lower. However, coming out, it would only take half an hour, it was all downhill then of course. If we were going too fast, we could use our kneepads to act like brakes, by pushing them against the wheels. We could also slow ourselves down by using our hands on the wheels, because they were full of oil, and we never got burnt. Everyone had their own sledge, they knew exactly which sledge was theirs – we would put a mark on them.

I was at Scaitcliffe when they broke into the workings at Hapton Valley Colliery, four and a half miles away from the Scaitcliffe shafts. We could hear the men at Hapton Valley talking on the other side of the coal, just before we broke through. We knew we were coming towards them, the management knew as well, because we had an air door ready to fit into where we broke through. This was so that it would not upset all the ventilation at the two pits. I remember one fellow who was called 'Sneck' for whatever reason, who shouted me over on the pit top one day, and he said to me 'Do me a favour Bill will you?' I said, 'OK, what do you want me to do?' He said, 'When we get down pit I am going to hitch a ride on one of the tubs, I am tired, I don't

Scaitcliffe Colliery, Accrington, was almost in the very heart of the town. (Ted Clarke)

feel right good today'. So he wanted me to take his sledge in for him, so he could ride out at the end of the shift. I got to the place where Sneck worked at the No.1 Junction near some air doors and waited for him. I could see his light at the back of me, he appeared to have stopped, so I went back to see what was the matter. When I got to him, he was dead – he had just stopped where he was, and died in the pit. I was petrified, I was only 17½. All I could do was ring the bell wire to indicate an emergency and wait for the fireman to come along and get the poor fellow out of the pit.

One of my most frightening moments at Scaitcliffe, though, was when we were working the pit out – getting at the old workings. The whole side of the coal and roof came away, a large rock, shaped like a bomb, wide at one end and tapering down at the other end – it came right down on top of us. My anchorman, while I was working the cutter then, was Charlie Silverwood, he caught the brunt of it, and he was crushed under the rock. The rock only caught me by the edge of it – but it put me off work for two years, I was in a plaster cast for seven months, I fractured my neck. If it was not for the coal cutter I would have been killed too; the rock fell on to the top of the machine, and even though it was only 18in high it stopped the stone from coming on

The last men out of Scaitcliffe Colliery, Accrington, on 29 June 1962, when the pit was closed down. From left to right: Len Holden, Tony O'Connor, Jack Duggan, Ted Barker, Albert Halliwell, Ken Dowling. (Ted Clarke)

top of me and killing me too. But I suppose I faired better than poor Charlie Silverwood – he were killed under it. There was no compulsory hard hats at this time, I used to wear a cloth cap, but a few did wear the hard hats made of compressed cardboard. The roof at Scaitcliffe was shaley, and you could always tell when it was about to come in because it used to start 'bitting': little pieces falling from the roof. We knew then to get off the face, because it was about to come in. The wood props we used at Scaitcliffe would warn us as well, they would creak and splinter to warn us that the weight was coming on. Some of the wood props looked like Chinese lanterns when they came out of the pit, all splintered and swelled in the middle. After the accident they would not let me back down the pit, I got £7,000 in compensation and invested in a café besides Joe Mort's in the town centre – I called it the 'Wayside Café'. I ran that for a couple of years, and then I went to William Blyth's Chemical Works, driving. I was there about five years. Then I went over to the Channel Islands with about ten mates of mine for about two years, before going to Australia. I was in Perth, Australia, for thirty years working on installing air conditioning and retired there, before coming back to England about ten years ago for family reasons.

At this point I asked Bill what he thought of his time in the coalmining industry.

It was great, I loved it, everyone looked after each other – they had to, as you know, Jack. I would do it all over again. Today I am virtually housebound, but I did enjoy a game of snooker and played pool a lot, but I just spend my time watching television now – I get plenty of company though, sometimes too much, especially the women!

If you are interested in purchasing other books published by The History Press
or in case you have difficulty finding any of our books in your local bookshop,
you can also place orders directly through our website

www.thehistorypress.co.uk